Ethereum Merge

Second-biggest Blockchain Has Completed From "proof Of Work" To "proof Of Stake"

(How To Make Intelligent Investments On Etherum And Ethereum 2.0 Merge)

Thomas Cooper

Published By **Andrew Zen**

Thomas Cooper

Ethereum Merge: Second-biggest Blockchain Has Completed From "proof Of Work" To "proof Of Stake" (How To Make Intelligent Investments On Etherum And Ethereum 2.0 Merge)

ISBN 978-1-77485-885-1

Legal & Disclaimer

TABLE OF CONTENTS

Introduction

If you've just realized the enormous opportunities with Ethereum and you're ready to join the party. This incredible cryptocurrency has lots of potential for both old and new enthusiasts of crypto. A myriad of innovative ideas have been developed using this Ethereum blockchain and many more are to come. Thus it is certain that Ethereum will be profitable over the long run.

If you're a novice perhaps you've encountered difficulties finding your way around the confusing internet of diverse information about Ethereum and other cryptocurrency. I faced this problem at the beginning of my journey. It took me some time to remove the seed from the shaft. Now, you've got all the data you need to understand the basics of Ethereum. I've ensured that.

This book will teach you the fundamentals details Blockchain as well as cryptocurrency with a particular focus on Ethereum.

Chapter 1: Introduction To Ethereum

What Is Ethereum?

In the event that Bitcoin (BTC) represents the alleged the future of currency, what exactly is Ethereum? If you're who is new to the cryptosphere this is a valid inquiry to make, as they'll surely find Ethereum as well as its own Ether (ETH) cryptocurrency as a close second in line with Bitcoin all over exchanges as well as in the media.

It's however not correct to think of Ethereum to compete directly with Bitcoin. It serves a variety of purposes as well as features and technology.

The core of the platform is that Ethereum is a global decentralized software platform that is powered through blockchain technology. It is mostly known by its currency of origin called ether, also known as ETH.

Ethereum can be utilized by anyone to build any type of digital technology that is secure. It is a token intended for use on the blockchain , however, it can be used by other participants as a way to pay for the activities performed through the blockchain.

Ethereum is expected to be scalable, programmableand secure and decentralized. It is the preferred blockchain preferred by companies and developers who are creating new technologies that rely on it that are changing the way that different industries operate and the way we live every day activities.

It supports natively smart contracts, which is the primary mechanism that powers decentralized apps.

A large portion of decentralized finance (DeFi) as well as other apps use smart contracts, which are paired with blockchain technology.

Many consider Ethereum as the next step for the internet. If centralized platforms such as the Apple App Store reflect Web 2.0 an open, user-powered, decentralized network such as Ethereum is Web 3.0. It is a "next-generation web" permits decentralized applications (DApps) as well as the decentralized finance (DeFi) as well as DEXs, or decentralized exchanges (DEXs) for example.

The most important facts

Ethereum is a platform that uses blockchain technology that is most famous because of its crypto currency ETH.

The blockchain technology behind Ethereum ensures that digital ledgers are secure and safe to be created and maintained.

Bitcoin and Ethereum share many similarities, but different goals and limitations for the long term.

Ethereum is moving towards operational systems that offers incentives for transactions to those who pledge their Ethereum.

Ethereum is the foundation for many of the upcoming technological developments.

What is Ethereum? Ethereum Function?

As with Bitcoin Like Bitcoin, Ethereum is similar to Bitcoin. Ethereum network runs on a vast number of computers because people act in the role of "nodes," rather than the central server. This makes the network highly resilient and decentralized to attack, and inconceivable to be destroyed in the event of a fall. If one machine fails it's not a problem because thousands of other machines remain in place to keep the network functioning.

Ethereum is a decentralized system that is run by a computer dubbed"the Ethereum Virtual Machine (EVM) (EVM). Each node runs an exact copy of this computer, meaning that all transactions must be authenticated to ensure that every user can upgrade their own copy.

Network interactions are also referred to as "transactions" that are recorded in blocks on the Ethereum blockchain. Miners check these blocks prior to placing them on the blockchain and acting as a history of transactions or an electronic ledger. The process of mining to validate transactions is referred to as proof of work (PoW) consensus method. Every block is assigned an unique 64-digit code that identifies it. Miners commit their computing power to identifying the code and proving the authenticity of it. Their computers are "proof" of their efforts and miners get paid with ETH for their work.

As with Bitcoin as well, the majority of Ethereum transactions are publicly available. Miners broadcast completed blocks to the entire network, which confirms the change and adding the blocks to every version of the ledger. Blocks that are confirmed cannot be altered with, thereby providing an exhaustive list of transactions made on the network.

If miners earn a reward for their hard work Where does the ETH come from? Every transaction is accompanied by the cost of "gas," which is paid by the person initiating the transaction. This cost is paid to the mining company that validates the transaction, encouraging the future mining process and providing security for the network. Gas functions as restrictor, which regulates the number of transactions users can perform in a single transaction. It also serves to stop network spam.

Since ETH is more than just a utility token, but one that has value it is a supply that is inexhaustible. Ether is constantly introduced to circulation through mining incentives. It can be used to pay for stakes and more if the system shifts to proof-of-stake (PoS) (PoS). In general, Ether will always be popular, which is why inflation shouldn't devalue the value of the asset above its usage.

For the majority of users, Ethereum gas prices may fluctuate quite high depending on the activity of the network. This is due to the fact that the block can only contain only a certain amount of gas, which fluctuates in accordance with transaction types and amounts. In the end the miners are more likely to prefer transactions that have the highest prices for gas, which means that they are fighting to validate the transactions first. This is a reason why prices are higher and higher, causing congestion on the network during peak times.

Congestion in networks is a major problem, however it's taken care of in Ethereum 2.0 A complete revamp that will be presented in a separate piece.

Engaging in Ethereum requires the use of bitcoin. Bitcoin is kept in the wallet. This wallet connects to DApps and acts as a way to access Ethereum. Ethereum network. Through it, anyone can purchase products or games, lend money and carry out all kinds of actions just as they would on the traditional internet. The only difference is that the traditional internet isn't free to users since they're giving their personal information to third parties. Websites run by centralized companies sell their data to make profits.

Cryptocurrency is a substitute for information, which means that users can browse and interact without revealing their identities. This implies DApp use is not discriminatory. In other words there is no way that a bank or loan DApp could deny someone access due to status or race. A middleman cannot stop what they consider an "suspicious deal." Users determine the actions they take and how they conduct it. This is the reason the reason why many people think Ethereum as Online 3.0 - the future of online interaction.

Vitalik Buterin, credited with the concept of Ethereum The Ethereum network, Vitalik Buterin, was the one to present an article in white to promote it in 2014.

Ethereum is a blockchain platform. Ethereum platform was launched at the end of 2015 in the year 2015 by Buterin as well as Joe Lubin, creators of the software company that developed blockchain technology called ConsenSys.

The developers of Ethereum were the first to analyze the capabilities of Blockchain technology, which goes beyond simply providing a secure digital payment system.

Since the introduction of Ethereum in the year 2000, the Ethereum cryptocurrency has expanded to become the second-largest cryptocurrency in terms of market value. It is only outnumbered by Bitcoin.

The proof-of-work procedure and the competitive reward system are two of the elements which have led to the development of large mining complexes , also known as mining farms. Mining farms are financed by richer companies and organizations that aim to regulate their mining processes.

What is Ethereum do?

Decentralized banking is perhaps the Ethereum network's biggest triumph. DApps that are able to perform different tasks within the ecosystem came from the year 2019 until 2020 and are increasing in popularity each day. The more DApps used in the future, the more it is likely that the Ethereum system will become utilized in the process. The Ethereum DeFi section is among the biggest one available and has been a success, with the DApps increasing the visibility of the platform in the past.

Artists, for instance generate millions of dollars from their artworks and putting it on the blockchain with nonfungible tokens, also known as NFTs. One might ask, why we should buy digital art instead of taking a picture of it? Collectors are demanding ownership, which is the reason. NFTs also provide proof of ownership and serve as secure storage. They're basically a complete package for collectors, and it's not difficult to grasp the appeal.

The same reason is why people would prefer"The original "Mona Lisa" instead of merely a copy regardless of whether a copy is indistinguishable from the original. NFTs also represent useful accessories and products in online games. They can also decorate their homes and their personalities with unique pieces from artists, providing a funding source for artists.

Developers have developed social networks that are not censored which allow users to pay each other for content. Games allow players to make investments in their assets then play to improve their value, and then sell for profit, making real value from their time playing. There are prediction systems that pay those who are able to make accurate predictions as well as freelancing platforms that don't require the majority of each payment.

The whole process is controlled by an autonomous system with the help of blockchain and smart contracts and DeFi giving customers greater control of their money than they have ever been.

Ethereum Mining

The process of creating blocks of transactions to become part of Ethereum's Ethereum blockchain is known as mining. Ethereum currently operates on a proof of work blockchain, but has shifted to a proof-of-stake (PoS) using Ethereum 2.0 for reasons of scalability and to be a more eco-friendly approach.

Ethereum miners are computers which run the software and utilize their time and energy to handle transactions and create blocks. The network members have to ensure that everyone is in agreement on the sequence of transactions in decentralized systems such as Ethereum. Miners help in this by creating blocks by solving complex computational puzzles and thereby protecting the network from unauthorized access.

Blockchain Technology

Ethereum like all cryptocurrency, utilizes blockchain technology. Imagine a huge block chain. The information contained in every block is added to each newly created block, with new information. In the entire network, the exact replica of the blockchain being distributed.

The blockchain is verified by a system of automated systems which reach an agreement on the authenticity of the transaction data. It is not possible to make changes to the blockchain before the system has reached an agreement. This makes it extremely secure.

Consensus is achieved through the process known as the consensus mechanism. Ethereum utilized the proof-of-work method which is a method by which a group of users runs software to verify that the encrypted number is legitimate. This is known as mining. The first person to prove the validity of the number is paid in the form of ether. Since mid-September 2022 Ethereum has officially switched to a proof of stake model that is faster and cheaper as well as eco-friendly than a proof of-work model. This change has led to an entirely different version of Ethereum called Ethereum 2.0.

Protocol for Proof-of Stake

At present, Ethereum employs the proof-of-work consensus system. In the future it will switch to a different consensus method known as proof-of stake, in which ETH holders stake a certain amount of their Ethereum. By staking their ether, they prevent it from being used in transactions. It serves as a reward and collateral to secure the mining right.

Mining will operate differently with this protocol as it doesn't require everyone on the network to battle for rewards. Instead this protocol is going to randomly select people with staked ether to validate transactions. The validators then get paid with ether in exchange in exchange for the effort they put into.

Wallets

Ethereum holders utilize wallets to keep their funds. A wallet is a virtual interface that allows you to access the ether that is that is stored in the blockchain. It has an address that is like an email address because it's where people send money, just as they would send emails.

Ether isn't stored inside your bank account. Your wallet is home to private keys that you use just like a password whenever you start a transaction. You will receive a private key for each ether account you have. This key is essential to access your network. This is the reason you get so many messages about protecting keys with various storage methods.

The History of ETHEREUM

Ethereum was not always the largest blockchain-based projects in the world. Vitalik Buterin co-created the project to address the shortcomings of Bitcoin. Buterin has released his Ethereum whitepaper in the year 2013 to explain smart contracts, which are automated non-changeable "if-then" statements which allow for the development of decentralized applications. Although DApp development existed before in the blockchain space, the platforms were not compatible. Buterin invented Ethereum to bring them together. For him, harmonising the ways DApps operate and interoperate was the sole way to maintain the interest of users.

So, Ethereum 1.0 was created. Think of it as the Apple App Store: one location that houses tens of thousands various programs, all bound by the same ruleset. This ruleset is embedded in the system and implemented by the independent with DApp developers being able to apply their own rules in DApps. There isn't any central authority such as Apple making and enforcing its rules. Instead, power lies at the disposal of the members of the community.

Of course, building an enterprise isn't cheap. That's why Buterin along with his co-foundersthe group of Gavin Wood, Jeffrey Wilcke, Charles Hoskinson, Mihai Alisie, Anthony Di Iorio, and Amir Chetrit -- held an open-to-the-public token sale to raise $18,439,086 for Ether which will be used to finance Ethereum's current as well as future development.

The consortium also launched an Ethereum Foundation Switzerland in order to preserve and grow the network. Then, Buterin announced that the foundation will operate as a non-profit organization which caused several co-founders to leave the organization.

In the course of time, developers came to Ethereum using their decentralized ideas. In 2016, the users formed The DAO, a democratic body that could vote on the future of Ethereum and made suggestions. The DAO was funded through a smart contract that did away with the need for a CEO to hold the power over Ethereum. In lieu, the majority of the Ethereum community is required to approve any changes before they can be implemented.

But, the whole thing was a disaster when an unidentified hacker stole $40 million worth of funds from the DAO's accounts due to a security flaw. To stop the theft, The DAO decided for "hard create a fork" Ethereum, diverging from the previous network and converting to the new protocol, basically going through a major software upgrade. The new fork kept the name Ethereum and the original network remains under the name of Ethereum Classic.

The most significant historical event that Ethereum has experienced has been the hard fork or split, of Ethereum in the form of Ethereum Classic. The year was 2016, and a small group of network members gained the majority of control over the Ethereum blockchain and stole over $50 million in Ethereum, which was paid for by a project called The DAO.

The success of the raid was attributed to the involvement of a third-party developer to the project. A majority members of the Ethereum community decided to fix the fraud by invalidating the original Ethereum blockchain and embracing the blockchain that has been corrected historical record.

But, a portion of the community decided to preserve an original copy of Ethereum blockchain. The unmodified version of Ethereum has been irrevocably split into Ethereum Classic (ETC). Ethereum Classic (ETC) (ETC).

Ethereum is vs. Bitcoin

Ethereum can be compared with Bitcoin. While both cryptocurrencies share numerous similarities, they do have certain important differences.

Ethereum is described by the group as "the world's most programmable blockchain," positioning itself as an electronic, programmable system that has numerous applications.The Bitcoin blockchain, by its own definition, was designed to facilitate Bitcoin, the bitcoin cryptocurrency.

It is believed that the Ethereum platform was established with the goal of leveraging blockchain technology to support a variety of applications. Bitcoin was developed solely to be an payment method.The maximum amount of bitcoins that could be put into circulation is 21 million.

Amount of ETH that can be generated is limitless, however the amount of time needed to process one block of ETH restricts the amount of ether that can be created every year.The amount of Ethereum coins that are in circulation is higher than 120 millions.

A major difference between Ethereum the Bitcoin network and Ethereum Bitcoin is how both networks handle the processing of transactions. These charges, also known in the context of gas within the Ethereum network are paid by members of Ethereum transactions. The costs that are associated with Bitcoin transactions are taken care of by the wider Bitcoin network.

Although Bitcoin has been the longest-running popular cryptocurrency However, the Ethereum community is aiming to further develop the reach of the project. Bitcoin is intended to be a digital form of money and serves this purpose fairly well. However, Bitcoin has some limitations. It's an PoW network that is experiencing difficulties in scaling, leading many to believe it's more a store of value, just like gold. Bitcoin is also regulated by a limit that is 21 million dollars making it more a part of that idea.

Ethereum is, on the other hand, plans to replace our current infrastructure. It is planning to automate a lot of procedures that require intermediaries like making use of an app store, as well as working alongside fund management companies. ETH is more of an interface with the internet than an option for money transfer, but it could be used for that as well.

Developers are able to build on Ethereum to create a distinct Ethereum-compatible token for every DApp known as the ERC-20 token. While it's not perfect however, it does mean that all Ethereum-based tokens can be technically interoperable. The Bitcoin network is only used to support Bitcoin.

One of the ways that Ethereum and Bitcoin have in common is the fact that both blockchain networks use huge quantities of energy. This is because both of these blockchains utilizes the proof-ofwork protocol. The proof-of-stake protocol uses less energy.

Future of Ethereum Future of Ethereum

Ethereum's move to the proof-of-stake protocol, which allows users to verify transactions and create new ETH using their existing Ethereum holdings and is an important update of Ethereum. Ethereum platform. Prior to this, the Ethereum platform was known as Eth2 the update is now known as the "consensus layer.

This upgrade is also expected to boost the capacity of it. Ethereum network to accommodate its expansion, which will assist in solving the chronic issues with network congestion which have pushed up the cost of gas.

Ethereum adoption continuesto grow, with the majority of it being the most prominent companies. In the year 2020, chipmaker Advanced Micro Devices (AMD) announced an alliance with ConsenSys to establish data centers that are built using Ethereum. Ethereum platform.

From 2015 onwards, Microsoft has had a collaboration with ConsenSys to create Ethereum Blockchain as a Service (EBaaS) technology that Microsoft uses on its Azure cloud-based platform.

It is said that the Ethereum blockchain has experienced increasing popularity in recent months because developers have utilized it to create a variety of decentralized finance projects as well as NFTs. The advent of new applications such as those -- which are among the first ones to be operate on a blockchain that is public has already created an enormous network effect according to the advocates and developers, as the increased activity draws increasing numbers of users to Ethereum.

But, fundamental questions remain over the possibility that Ethereum is behind schedule and has an array of technical advancements, is capable of competing with faster-moving competitors and whether a consensus on its future role will be able to emerge as the crypto market expands.

However, some investors such as Garg caution that, given Ethereum's long-term importance that the cryptocurrency market could be in the process of reversing in the near future, with Bitcoin rising back to the undisputed top spot.

Web3

Web3 is an idea that is still being developed, but it is widely believed that it is powered by Ethereum because a large number of applications that are being developed utilize it.

Utilization in Gaming

Ethereum has also been integrated into virtual reality and gaming. Decentraland is an online realm which makes use of blockchain technology called the Ethereum blockchain to protect the items inside the virtual world. Avatars, land, wearables as well as buildings and environments are all tokenized on the blockchain system to establish ownership.

Axie Infinity is a different game that makes use of blockchain technology. It also has its own cryptocurrency known as Smooth Love Potion (SLP) that is used to reward players and for transactions in the game.

Non-Fungible Tokens

NFTs, also known as non-fungible tokens (NFTs) have gained prominence in 2021. NFTs are digital tokens made using Ethereum.

In general, tokenization grants an individual digital asset a unique digital token which identifies it and is stored in the Blockchain.

This is proof of ownership as the encrypted data contains the wallet address of the owner. This is why the NFT can be sold or traded. This is seen as an transaction through the blockchain. The transaction is confirmed by the network, and ownership transfers.

NFTs are currently being developed for various kinds of assets. For example, fans of sports can purchase a token of their favorite sports team, often called fan tokens athletes, which could be used like trading cards. A few among these NFTs are images that resemble the appearance of a trading card, while others are video clips of a significant or memorable moment in the career of an athlete.

The programs you use in the metaverse like your wallet, DApp, or the virtual world and the buildings that you visit, are most likely to be built on Ethereum.

Development of DAOs Development of DAOs

Decentralized Autonomous Organisations (DAOs) that provide an approach to collaboratively making decisions on an unconnected network they are currently being designed.

Imagine, for instance, you set up a venture capital fund , and did fund-raising to raise money, but you'd like the decision-making process to be decentralized, and distributions to be automated and transparent.

A DAO could make use of smart contracts and other applications to collect votes of the fund's members and invest in ventures on the basis of the majority of group votes. It will then divide any profits. The transactions would be seen by all parties and there is no involvement of third parties in the handling of any money.

The role that cryptocurrency could be playing in the near future isn't clear. Yet, Ethereum appears to have an important, future function in corporate and personal finance, as well as in the various aspects of our daily lives.

How can I Buy Ethereum?

You can't purchase cryptocurrency from banks or an online brokerage such as Vanguard and Fidelity. Instead, you'll have utilize an exchange platform for trading in cryptocurrency. There are many cryptocurrency exchanges to choose from with a range of options from simple to more complex dashboards designed for traders with advanced skills. Different platforms come with different pricing as well as security and other options that is why doing some investigation prior to signing up for a new account is a great option.

Investors can utilize any of the numerous crypto exchange services to purchase and sell the ether. Ethereum is accessible through dedicated cryptocurrency exchanges like Coinbase, Kraken, Gemini, Binance, and brokerages such as Robinhood.

How to buy Ethereum

To establish an account at an exchange for crypto it is likely that you'll have to provide certain personal details and be able to prove your identity. You'll then be able to add funds to your account by linking to your debit or bank account. The fees will likely differ depending on the payment method you select.

A successful account opening does not necessarily mean you've accumulated Ethereum or any other cryptocurrency. As is the case with any investment account you shouldn't allow your funds that aren't invested to remain inactive. At this point you need to purchase Ethereum for investing.

You'll be able trade with your United States dollars for Ethereum when your account is fully refilled. Simply type in the dollar amount you'd like to trade to Ethereum. Based on the price of Ethereum and how much you'd like to purchase, you'll most likely be purchasing shares of one Ethereum currency. Your purchase will be shown as a percentage of the entire ether coin.

It is easier to transfer your cryptocurrency investments in your account for exchange even if you have only just a small amount. However, if you'd like transfer your investments to a secure storage place the digital wallet could offer additional security. There are many types of digital wallets, each having different levels of security, like a paper-based wallet or mobile.

How does Ethereum Earn Money?

Ethereum isn't a central organisation that earns money. Validators and miners who take part in running in the Ethereum network, mostly through mining, receive ETH reward for their efforts.

Are Ethereum an Investment that's worth it?

Like all investments decision, the best answer will depend on your financial needs as well as your goals and the risk you are willing to take. The cryptocurrency ETH isn't stable which puts your capital at risk. But, it's definitely worth a look due to the many technological advancements that Ethereum employs could play larger roles in the society of the near future.

Is Ethereum a cryptocurrency?

The Ethereum platform is its own cryptocurrency that is also known as ether or ETH. Ethereum is a blockchain technology platform that allows for an array of distributed applications (dApps) that include cryptocurrencies. It is the ETH coin is usually referred to as Ethereum however the distinction is Ethereum is a platform powered by blockchain and ether is the cryptocurrency.

Is it possible to convert cryptocurrency into cash?

Yes. Investors who own the cryptocurrency ETH can make use of online exchanges like Coinbase, Kraken, and Gemini for this procedure. Create an account with the exchange, join an account with a bank, and transfer ETH into the bank account using the Ethereum wallet. Make an order with the exchange for selling ETH. Once sold make sure you pay your U.S. dollar proceeds to the bank account linked to.

The investment in cryptocurrency and Initial coin offerings (ICOs) are extremely risky and speculation This article is not an endorsement or recommendation by Investopedia or the author to invest in cryptos or ICOs. Because every person's situation is individual, a expert should always be consulted prior to making any financial decision. Investopedia does not make any representations or warranties regarding the timeliness or accuracy of the information in this article. In the year that this article was published the author is the owner of Bitcoin and Ripple.

Should you invest in Ethereum?

Ethereum is the second-highest valued cryptocurrency in terms of market capitalization and is considered to be the gold to Bitcoin's silver. Like all investments it's possible that the higher risk of Ethereum translates with higher rewards. It's not 2009 anymore: Ethereum has moved past the stage of proof-of-concept and is now the perfect time for those who haven't yet explored the possibility of this asset class to start.

Due to the volatility and uncertainty of the cryptocurrency market, prior to investing a substantial portion of your retirement savings to Ethereum as well as any of the other cryptos, you should do your own research. It could be worthwhile to consider it as an aggressive growth strategy in a diverse portfolio. Be sure to invest more than you are able to lose.

The advantages of Ethereum

In addition to decentralization and privacy, Ethereum also has various advantages, including an absence of the censorship. If, for instance, someone tweets something that is offensive, Twitter can choose to remove it and penalize the user. But, with an Ethereum-based social network it can only happen when the community decides for it. This way, people with differing opinions can talk according to their own preferences and the community can decide what is appropriate and what should not be said.

Community rules also stop malicious actors from being able to take over. Someone with malicious intent will need to manage 51 percent of the network in order to change their behavior that is almost impossible in many instances. It's much more secure than a server that could be cut into.

Smart contracts are another option that automate a lot of the actions performed by central authorities using the traditional internet. For a freelancer such platforms as Upwork is required to make use of the platform to locate clients and create payment agreements.

Upwork's business model uses an amount of the contract in order to reimburse its workers, servers expenses and so on. In Web 3.0, a client could simply create a smart contract that says "If it is submitted by certain times, cash will flow." The terms and conditions are encoded in the contract, and can't be altered by any the other once it is it is written.

It's becoming more simple than ever to get Ether. Companies such as PayPal along with its Venmo subsidiary let you buy cryptocurrency using fiat currency directly within the app. With the millions of users on every platform, they're likely to join sooner than later.

The disadvantages of Ethereum

Although it may sound like the ideal platform, Ethereum has some key problems that must be addressed.

In the first place, scalability. Buterin saw Ethereum as the way the internet is today and has millions of users interfacing simultaneously. Because of PoW's PoW consensus algorithm the interaction can be restricted by block validation time and gas charges. In addition, decentralization is an obstacle. Centralized entities, such as Visa has control of everything and has perfected the payment procedure.

The second is accessibility. At the date of this article, Ethereum is expensive to create and is difficult to communicate with those who aren't familiar with the technology. Certain applications require unique wallets, meaning that users must transfer the ETH they have in their current wallet to the desired wallet. This is a waste of time for those who are a part of our financial ecosystem, and not particularly user-friendly.

It's true that PayPal has added crypto-related support but customers can't accomplish much other than keeping it there. PayPal needs to be integrated to DeFi or DApps to make it more accessible in a significant way.

The Ethereum platform has written documentation that is well-written as a way to increase the number of users. However, the actual process of using Ethereum is a process that needs to be streamlined. Understanding the blockchain is different from actually using it.

Chapter 2: What Is Ethereum?

Ethereum can be described as an open programming stage consisting of a distributed group of virtual machines that are dependent on blockchain innovations. It is a platform for developers who wish to build and run distributed applications.

Developers are able to send distributed or decentralized applications (Dapps) that function without extortion, controlor obstruction from an outsider and personal space. Ethereum is an electronic stage as well as an programming language that suddenly increases the demand for blockchain.

It allows engineers to create and distribute a variety of programs. The Ethereum virtual machine organization is advanced enough to execute applications that are decentralized when specific conditions are fulfilled, such as the execution of contracts. Ethereum is a decentralized blockchain which utilizes cryptography technology to execute, store and protect these contracts.

Ether The Ether token is a part of the Ethereum organisation, there is an Ethereum token, also known as. Any applications that spike in popularity for the Ethereum company and its blockchain use the Ethereum token. Ether is an cryptograph token that is the foundation of any application that runs on Ethereum. It is often portrayed as a way to explore the Ethereum stage, so anyone who is thinking of Dapps or distributed applications must get ether.

Ethereum is used by the company to achieve two major goals. One is to be used around as a currency or vehicle for trade and the other is to enable applications to run that run on Ethereum. Ethereum blockchain. Ether is a crypto currency used by developers to pay for services and exchanges on the Ethereum network.

Exploring Ethereum

Ethereum is an option to use with any other function. One of the most current tasks that is the largest activity on the Ethereum stage is the coordination between

Microsoft as well as ConsenSys. This company has observed ConsenSys provide Ethereum

Blockchain administration on Microsoft Azure. This allows engineers as well as customers to get access to the cloud-based blockchain development with one click.

Ethereum functions as both an application store decentralized to the Internet because it allows another type of use. It is not owned by anyone however it's not completely accessible to all users. Everyone requires ether to run their programs. Ethers that are used to run the framework are similar to money or resources. They are the fuel for the apps within the network.

Ethereum offers what is referred to by"the Ethereum Virtual Machine. It is referred to as an decentralized Turing fully-fledged machine. It is the machine which executes content or application software on its massive network of hubs for public use. Every hub has to download and run this virtual computer.

When developers run their designs on the framework and have to pay "gas" by using Ethers. Gas refers to the internal evaluation framework that is applied directly to the framework. It distributes resources and prevents spam. Hubs that participate in the program are paid using the ethers.

Origins of Ethereum

Vitalik Buterin is a developer of digital money and examiner master. He was an engineer for another blockchain company, Bitcoin. The reason he created Ethereum was to provide an opportunity to develop applications that are decentralized. As stated by Vitalik, Bitcoin needed a prearranging language that could be used to create applications. Thus, he used Bitcoin to develop Ethereum.

In the latter part of 2013, he wrote an article that proposed Ethereum. Shortly after the release of the whitepaper, funds for its growth were raised via an internet-based crowdsale. The sale took place between July and August 2014 and raised vital assets. Customers bought 60 million ethers and around 12 million was set aside to developers and software engineers.

Ethereum was launched in July of this year. Following its launch, 11.9 million ethers were released for crowdsale. They account for about 13% of the total supply of ethers. The framework is populated by diggers who are working to make new Ethereum. To accomplish their tasks they purchase ethers that are a kind of cryptocurrency.

DAO -- Decentralized Autonomous Organization

In the year 2016, Ethereum split into two forks. The split was brought on due to the collapse of a function called DAO. The one fork is known by the name of Ethereum (ETH) and another is known being Ethereum Classic (ETC). The most popular method of dividing of the fork is referred to as the hard fork. It is the Ethereum hard fork has been a source of an argument between the two emerging systems.

The DAO incident is an organization that was created in the year 2016. Decentralized Autonomous Organization, or DAO made up of smart agreements that were developed using Ethereum. These agreements were launched during a crowdsale that collected $150 million that was expected to help the project. However during June, the framework was compromised and hoodlums found a way to break into the system and took the $50 million worth of

38

electronic currency. This burglary resulted in the hard fork of the system.

Ethereum virtual machine Ethereum virtual machine

EVM which is also known as Ethereum virtual machine is a reference to the conditions that are required for contracting smartly. The EVM is more precisely described in the Ethereum Yellow Paper composed by Gavin Wood. The current design framework is distinct from the rest of the framework, including documents, organization, and any other procedures.

The computers within the company known as hubs, must run EMV. Virtual machines run in various advanced programming languages like Java, C++, Ruby, and Python as well as other languages. Smart contracts

A shrewd contract refers to an PC program that runs on the Ethereum stage that is able to work to facilitate the trading of important items that have a value like content, property offers, or cash. The purpose behind shrewd agreements is to offer the ability to trade, which allows exchanges between two experts. These agreements are designed to safeguard and validate exchanges and also avoid plots or oversight and risks between the two parties.

The Ethereum blockchain, smart contracts function as the decentralized software that is stored within it to be executed in the future , by Ethereum virtual machine. These contracts are executed using a variety of programming languages and are funded with ethers or gas.

The amazing agreements are publicly available on each network hub on the Ethereum blockchain. The most important test is slower speeds are embraced by the framework on account that every hub in the framework must figure out each of the clever agreements in real time.

There are numerous blockchains that are able to efficiently cycle code. The problem is that they are limited in a broad array of ways. Ethereum is, be that as it is, does not have to face some of these limitations in its operations. It provides engineers with the ability to create a variety of applications that are capable of achieving astonishing feats.

The benefits of Ethereum

Ethereum offers a variety of benefits to its users. It is essentially empowering software engineers to design and deliver a range of applications that are decentralized. Dapps, also known as decentralized applications, are developed using software that runs with the blockchain, so they're not controlled by any authority or person.

The framework allows any unifying application to decentralize. There are many administrations that are currently uni-directional which can be decentralized. This includes races on the board and land title vaults. advance and loans from monetary institutions such as auto enrollment, and many more.

One of the major advantages that comes with one of the major advantages offered by Ethereum platform is that it's indestructible and nobody can alter the information , information or code within the framework. Another benefit is that the entire framework is designed with care and isn't susceptible to being destroyed. The applications designed to be used on the Ethereum stage are based on a rule of law within the network.

The system is extremely secure. It's built using the most up-to-date cryptograph technology which is secure and free of any crucial issue of failure or dissatisfaction. A secure environment is ideal for running programs that are protected from attacks of extortion and hacking. In addition, the framework experiences virtually no personal interruptions. The framework's applications can't be disabled by anyone and are never offline.

Basics of cryptocurrency

The use of cryptocurrency makes it extremely simple to transfer and receive assets within the company. It's much less expensive and quicker to utilize cryptocurrencies instead of traditional methods of money transfer like wire transfers, and so on. Cost savings is one of the primary benefits of this. For instance, financial institutions like banks often charge a large amounts of cash to transfer money, especially international wire transfers.

Another benefit of digital currencies is the fact that individual information is not necessary for the execution. Anyone can use digital currencies without having to discover their nature. Assets can be transferred to recipients without needing the existence of a financial balance. Additionally the digital money stages won't offer any assistance to anyone and don't separate. Anyone can use the modern financial forms, and no account is able to be closed or the assets taken. This is a clear distinction between the kind of banking institutions that are capable of performing.

Blockchain

The blockchain is essential in the development of cryptographic currencies. It is a crucial component of any digital currency since it records the exchanges within the company. The online record is updated regularly and customers can view the exchanges on their devices. This kind level of transparency, decentralization, and decentralization are unimaginable for the financial world. It is expected that in the near future, monetary foundations and other associations will rely advantage of a more open framework, most likely with a blockchain-based framework that will provide greater transparency in their business operations.

A lot of experts believe that the blockchain has potential in a variety of areas. It could be used in areas such as dealing with monetary exchanges, storing exchanges between securities and many other areas. Large financial institutions like JP Morgan are of the opinion that a blockchain technology could help companies save billions of dollars in cost of transactions.

Chapter 3: What Is Cryptocurrency?

A cryptographic currency can be described as a type of digital cash used online. Digital money makes use of cryptography technology to create cash and exchanges. The money is virtual currency that doesn't have in a distinct form. Exchanges are not known and are created from the technology used.

Cryptocurrencies operate with a dispersed records framework that allows all users to view exchanges. There is no central control and no authority or management is in charge of the money. The electronic currency is decentralized and is among its fundamental characteristics.

The development of cryptography grew from the desire for secure communication. The present day use of cryptography incorporates elements of software engineering and numerical theories to offer an effective method of communicating. By using cryptograph technology, information is encrypted and transmitted in a secure and secure manner that is difficult to break. This technology is used to transmit data securely and also for information security and verification.

Additionally, it is used to secure correspondence and assets online.

Prior to 2009, there were plans to create an decentralized money using an organization that disseminated and cryptograph technology. Finally, in 2009 the first of these moneys came into existence. It is Bitcoin that is currently the most well-known cryptocurrency around the globe today.

A single of the unique features that digital currencies possess is their inherent nature. Digital currencies aren't governed by states or other substances brought together. This means they are free from the shackles of government or any interference.

The cryptocurrency is not made of paper or stamped as coins and are not backed by a large amount of resources, such as gold. In the end, they occur in a pattern called mining. Mining companies invest their time and energy in tackling difficult

Numerical riddles and, when they are solved, digital forms of money are brought to be created.

What is the function of cryptocurrencies?

The nature of cryptocurrencies is not known to the public. The clients don't need be able to add their names to exchanges and there's no need for ledgers. The system is decentralized, which means that clients can transfer money and receive cash, or even make installment payments by way of a public record that is disseminated that is referred to in the field of Blockchain.

The blockchain is a publicly accessible record of all transactions recorded. It isn't an unchanging record because it is constantly updated. Every client can access the blockchain and are able to see an entire history of transactions. The disseminated record system also functions as a shared platform where users are able to perform their transactions directly between themselves, without the necessity of an intermediary.

Anyone in any part of the world is permitted to use digital currencies. There is no restriction based on race, sexual orientation and ideology, or geographical the identity of the person. There are stages and trades online that allow cryptographic forms of currency are traded or bought with other currencies.

Each client or record is linked to the rest by a complex set of numerical requirements. This ensures that each exchange is authentic and exact. The task of confirming the legitimacy of exchanges, validating them, and putting them on the blockchain is performed by excavators. Excavators are typically computer programmers and software engineers. They are paid for their efforts through costs of exchange as well as the cash they extract.

The use of virtual currency makes the movement of money from one client and moving on to the next beneficial, swift and easy. To transfer digital funds starting with one client and moving to the next, you must make use of two specific keys. Because virtual currency forms are distinct they are addressed with alphanumeric characters that are referred to by the term keys. One key is a private key, while the one that is public keys. Keys are also encoded they are to be used for authentication purposes.

When assets are moved beginning at one client, and moving to the next one, there is no expense for exchange. Every now and then, there is an expense that is not actually to be incurred. The cost is negligible and is usually utilized to pay excavators, who are the experts who verify exchanges and then add these on the blockchain.

Challenges

There are issues with the cryptographic form of money. One of the biggest issues stems from the sophisticated concept of these standards for monetary transactions. Since they are digital and exist only on the internet, users might lose all their assets in the event of a PC crash, in the event there was no protection. There are businesses who have lost huge amounts millions of dollars in cryptographic versions of money due to the crash of a computer. The backup of data running on the cryptographic money framework is crucial.

Volatility

The cryptocurrency market is extremely volatile. The price or value often is contingent on supply and demand. Financial backers may lose a large amount of money on the off chance they purchase huge quantities of digital currencies at the cost of a plunge just a few minutes after.

The uncertainty of digital currencies is usually due to the speculative nature of the clients. A lot of people purchase electronic monetary standard with the smallest amount of but they are rushing to sell it to make gains. To reduce this risk the risk, customers must find out how to buy the cash and hold it for a time.

Hacking

Hacking is a concern that is not unique to digital currencies. There have been several major attacks that have affected popular digital currencies. For instance, Bitcoin, the most well-known cash, was compromised and financial information stolen over 40 times. Some of these robberies involved losses of over $1,000,000. Fortunately, the flaws were identified and rectified.

The players are advised to be cautious and secure that their theories are confirmed. For instance, they can make use of a reliable secret phrase, or take on various marks, and store digital forms of money that are not linked. These simple advances could lead to digital currencies secure from hackers and cyber criminals. Despite all the challenges there are many who believe that cryptocurrency is going to be around for a while and be in the near future be accepted by the world at large. They protect the value of money they are not subject to supervision, and are more easy to transfer between nations and can be used to trade.

Summary of cryptocurrency

In the vast majority, there are over 700 digital currencies around the planet right now. The wealth of computerized financial types were developed after Bitcoin which is the cryptocurrency that underlies it.

The cryptocurrency that is not Bitcoin are all in the same way as altcoins.

There are transactions where customary clients and financial backers can buy digital currencies. Prior to investing resources into digital currency, customers are advised to exercise due decision. The majority of these are not sound and some are even completely unsustainable which means financial backers are losing money.

Instead of relying on speculation financial backing should take a stand for examination, tolerance, and a lot of caution.

The top altcoins on the market currently include Ethereum along with Litecoin. It is essential to keep in mind that there are customers who are unsure that the new financial backers will not steal their money.

Chapter 4: The Blockchain

The cryptocurrency industry requires structures and designs that allow their operation safely and effectively. At the heart of every profitable digital currency will be the blockchain. It is a crucial step that ensures that all transactions are secured and decentralized. What is the blockchain?

The blockchain is described as a computerized, decentralized public record in which all transactions in cryptographic currency are documented. When exchanges occur they need to be listed as valid, verified and confirmed. Each exchange is recorded in a single square. The square, after being confirmed it is then included in the chain.

The blockchain therefore is an unrivalled public record which continues to evolve continuously. Customers can access the blockchain anytime and make sure that they are able to confirm transactions. Blocks on the blockchain are transferred and linked with the latest cryptograph technology.

They are put in sequential request meaning that when a square is accepted, that it will be added to the frame. The square is called a hash pointer. The hash is focused on the previous block. The square is filled with exchange information as well as a time stamp.

Blockchains are designed to prevent any modification of data. This is important because diggers affirm the digital currency exchange prior to incorporating their data into the system. To keep pace with

confidence in the framework the framework is a must, and the information should remain constant.

In its most simple form the blockchain is an ongoing computerized record that has small pieces of data being added periodically. In addition, the blockchain could be utilized as a distributed record. This is where it might be used by the digital currency. In these instances the blockchain functions as a joint organization. When block is added the convention that has been endorsed, the information that they hold cannot be altered. If change does occur, it will affect each and every square that follows, and that's a decision that requires the assistance of an

The majority of clients are part of the majority of clients within the.

The first blockchain was created and developed to work with bitcoin, which was the initial cryptocurrency. Bitcoin. The technology used for its use is known by the term DLT also known as distributed ledger technology. Blockchain technology is being utilized in diverse sectors. In the main, it is utilized for processing transactions, entering them into and verifying transactions in cryptocurrency systems.

In addition to confirming exchanges Blockchains can also be modified so that it acknowledges every report and all data entered is an irrevocable record that is able to be verified and verified by users, but cannot be altered. If modifications be required, all the participants will need to agree, not just one centralized entity.

Blockchains provide an impressive illustration of the appropriated registration frameworks that are extremely secure according to design. The unique features of the blockchain make the blockchain a great platform to keep reliable and unalterable records like dental and medical records as well as firearms and motor vehicle registration information and identity management, as well as records management and land titles and many more.

Blockchain technology is the basis for its development.

Blockchain is a distributed entity that operates on the Internet. It is described as an advanced dispersed record , with the ability to circulate data sets.

Decentralization is one of the key components of this technology. Another unique aspect of the blockchain is the data set, which includes every square, data and other information is accessible across multiple computers within the network.

The blockchain network is comprised of computers, referred to as hubs. Hubs are interconnected to form what is known as a shared company. This kind of organization takes out the requirement for a PC server that is brought together.

Many organizations across the world employ integrated frameworks, which include an information set and server PC. The server, in the end, transforms into a useful goal to

Programmers and pernicious attackers. Blockchain technology is difficult for programmers to gain access in due to the fact that each and every piece of data is encrypted, decentralized and accessible to all users.

Each time there's an activity in the network, such as an order, payment, etc. the details of the transaction must be documented on all nodes or computers, in the network. In turn, all employees of the company, often referred to by the name of clients are granted access to the refreshed information. It isn't deleted or altered by anyone. At the end of the day, there's one record that has precise data, instead of several records that have conflicting data.

A brief overview of the history of blockchain

The blockchain first came into features in 2009, when Bitcoin came into existence. It is typically found in the bitcoin code. In the year Satoshi Nakamoto delivered a white paper describing the world's first digital currency and the world was able to observe the blockchain to its application.

Satoshi Nakamoto, a pen name used by the creators of bitcoin provided the code to the general public via the Internet. Satoshi further mined the underlying bitcoins, or digital currencies, to stage. The square on which bitcoins were mining was described as the first square. When they mined Bitcoins for the very first time, they introduced the world with the first cryptocurrency.

The world's attention was drawn to bitcoins especially after 2013. Everyone needed bitcoins, and thus the interest grew. The large amount of people who purchased bitcoins or sold them executed transactions through the blockchain. It is the most fundamental component of bitcoin.

Since since then, there have been a growing number of cryptocurrency and they all have their roots in the blockchain. Many different associations and foundations have expressed interest in blockchain. In short, the blockchain can be described as an electronic system that has applications with DLT or disseminated records technologies.

Tech experts affirm that this is the biggest technological advancement since the time of its creation.

of the of the internet.

Digital signatures

Every transaction on the blockchain is accompanied by the computerized signature. The signature is entered using private key cryptography. This technique makes use of two keys. One is open to the public, and the second is private, to cover transactions. Everyone in the company or all hubs within the blockchain process this information and can on the lines verify, confirm or verify any transaction.

However, only the data that is dependent on the public key are accessible to all employees of the company. The private key is accessible to the person who received it in order to understand the message and actually take a look at the subtleties. The keys can also be used to verify the identity of a person.

Blocks of information

The name blockchain comes because of the various squares that it is composed of. One square holds data related to different exchanges. The square also contains additional information. the square too. The data includes a period stamp and a reference to a previous or parent block, proof of work, and a header. The header is comprised of 3 pieces of information, including the time stamp as well as evidence of work and details of exchange.

Mining of cryptocurrency

Every digital currency must be mined to create real. Mining can be described as the process of creating new digital currencies, or the most popular method of adding squares to the blockchain.

Different types of digital currency come with different speeds for mining. On the Bitcoin level, Bitcoin takes about 10 mins to extract a square. On Ethereum the process takes shorter time. Therefore, it takes 10 minutes or less complete an Bitcoin transaction.

Every exchange that is included in the blockchain has to be verified. Mining is the method by which the process of approval is improved. Diggers, who are typically an engineer in software or another expert, must tackle a difficult problem in a numerical way. This process requires massive calculating power, and it burns through both time and power.

The complexity of the numbers continues to increase and the PC needs to be upgraded with power. Diggers are paid some of the digital currency they mine, and also are eligible to share some of the expenses for exchange that are incurred by the platform.

The uses of blockchain

The possibilities of transforming different areas of the economy by leveraging blockchain technology is huge. This is due to the fact that the blockchain functions as an innovation for establishment that is able to be modified into a more recognizable form, accepted, and adapted to make it a more specific application. It may even be able to design an institution for financial and social structures that are embraced by all over the world.

The method used by the blockchain to answer questions about frameworks is unique to the standard method that actions plans are altered typically to provide more affordable arrangements that overpower companies with their traditional models. Despite this rapid innovation and validation of ideas There aren't many structures that are based on blockchain. This is due to the fact that blockchain technology is still an emerging technology and lots of products actually are being created.

Blockchain applications could be problematic in their nature and may alter the way that current models operate. Blockchains are able to be integrated in a variety of areas for businesses to be able to capitalize advantage of current practices for managing payments and exchanges. Blockchains remove the need for delegated meetings and eliminate mediators. Blockchains also remove the need for professional organisations. They instead provide conventions for cycles such as advanced exchanges and the use of virtual currencies. The multitude of options available allow companies to cut costs and cut costs on every case expensive procedures. The digital ledger feature of blockchain is a way to eliminate risk like theft, fraud and other risks. Utilizing apps that use this feature can eliminate the need for a trust specialist cooperative. This means that less resources are able to be used in the resolution of problems. Additionally, the risk of extortion and theft inside the structure will be greatly diminished and the risk of foundational dangers eliminated. The frameworks that were slow and manual will become computerized, which will save time, effort and cutting the cost of maintenance considerably. It is possible that the results of this new technology must be apparent in the charge selection process, and risking the executive and transportation.

Certain applications of the blockchain

Blockchain currently has several significant applications. In any event, there are numerous significant companies which are working with blockchain-based applications. They're geared toward providing the best kinds of support which reduce costs, speed up processing times, and eliminate fraud and theft.

As a distributed ledger system for cryptocurrency. The major cryptocurrency platforms like Bitcoin, Dash, NXT and Litecoin all work on blockchain.

As a distributed registry , such as Factom which offers an unalterable permanent, secure and indestructible record-keeping system. It is an uncentralized and distributed protocol.

A decentralized messaging service like Gems

As a cloud storage distributed service, such as Sia and

Storj and as an uncentralized system of polling similar to Tezos

Future applications and development of the blockchain

There are a variety of frameworks and projects being developed and carried out in different areas. For instance, the protection area is comprised of three distinct conveyance methods that include miniature protection sharing protection, the parametric protection. Each one of them is modified to work with the Blockchain Record System.

There is a web-based or advanced democratic, which is a web-based. This type of voting

Framework is based on blockchain and ensures a solid, accurate and unshakeable democratic integrity that can't be affected. Major banks are currently looking into strategies they can use to benefit from blockchain technology. If they are successful banks can implement it, they will cut back on administrative center operations as well as reduce the time spent handling exchange transactions and reduce the cost of operations.

Private and public processes

Distribution of books, stockpiling of information land enlistment, and the unique, distinctive craftsmanship of proof are mostly processes that are likely to be nurtured by the blockchain. Some of the biggest financial institutions, such as UBS have developed new computer labs that allow researchers and software engineers are able to conduct more research about how blockchain technology can be utilized to improve the efficiency of financial services industry.

In Scandinavian countries such as Norway and Sweden there is progress ahead of blockchain-based initiatives to manage libraries of land. If it is executed correctly the project will be easy to implement in relation to land issues and both parties will benefit from the speedy land transaction processes. Other European countries are exploring various ways to build property libraries based on the same platform.

Are smart contracts a good thing?

A great agreement is simply a PC program that is executed using the blockchain. Great agreements needn't be concerned with any human interaction and are able to be executed in varying degrees or entirely. These agreements be, for instance, executed electronic escrows. Escrow is a process which holds the assets of the buyer to be transferred or transferred to a vendor upon the transfer of particular concurred managements. Smart agreements can create an escrow that is computerized upon fulfillment of specific conditions.

Ethereum is a renowned platform that operates in the field of blockchain. One of its main tasks, Ethereum Solidity, is an intelligent contract that can be executed in the theater. The script is altered and implemented using a programming language called Turin Complete which has the ability to execute incredible contracts.

A stunning example of the application of smart contracts on blockchain can be seen in the music industry. A music cassette or CD is made by music makers, DJs, or even studio. The tape is then altered in a stunning agreement, and then is added to the blockchain. Every time music is played, each craftsman is paid accordingly using an eminent digital currency. A comparable smart contract was successfully executed by Deadly Buda, a US DJ from the United States, Deadly Buda.

The accounting and blockchain industries

The four major bookkeeping firms , including Ernst and Young have tracked the ways to incorporate blockchain into their processes. Consider for instance Ernst and Young. The representatives of the firm in Switzerland have cryptographic cash wallets and an Bitcoin ATM inside their Swiss offices. Other major bookkeeping companies such as Deloitte, KPMG and PWC are mostly working with some sort of blockchain, which is currently under test. It is widely believed that shortly, all these bookkeeping firms will be equipped with their own sophisticated software and offer more efficient bookkeeping solutions that are able to be cooked and tailored to each customer, making sure they receive speedier, better, and more accurate bookkeeping that is efficient and solid way.

The benefits of the blockchain

Industries across the globe are generally transformed by the blockchain technology. Legislative bodies, associations installment stages, and financial bases are all transferred to the new platform.

Blockchain has a major impact on how capacity management is conducted and the way associations function.

The reason why blockchains are thought to be so significant ?

In our current age the monetary, political general and legitimate frameworks are based on agreements and exchanges to work efficiently. Agreements and exchanges just like the documents kept are essential for the management of information, resources, and various assets. The truth is that the vast majority of cycles are represented by agreements and exchanges. The method by which these are stored and archived is completely outdated. The basic frameworks, apparatuses and models that are expected to store information are not unique. Blockchain is the place that these numerous challenges can be solved. Many companies across the world are seeking to implement the blockchain technology so that they can increase their productivity and benefit from the massive potential benefits it can bring.

Blockchain eliminates intermediaries such as lawyers and bankers.

The majority of jobs in the modern world require the supervision of brokers and other delegated. For instance banks act as delegates between leasers and customers, while attorneys act as agents between clients and law enforcement. A large amount of mediators could be eliminated and let individuals from the general population can access the services they require quickly. Through the use of blockchain technology businesses, organizations, and even individuals will be able to connect and interact with each other in a direct way.

The same thing is happening in bitcoin. Bitcoin is among the most reliable digital money forms that exists. It is based on a decentralized, circulating blockchain. Bitcoin simply connects buyers, dealers, venders and customers without the need for middlemen, such as banks. Blockchain-based exchanges that are shared are more efficient, cheaper and are valid. The exchanges recorded are stored on blockchain and the transaction data is available for verification.

Blockchain networks are a blockchain-based system where every single exchange as well as every cycle is documented. It is possible for clients to track any back installments or exchanges made to the client or an association. Each exchange, after being affirmed by the client, is confirmed and comes with the mark of returning. This is an effective method of establishing a business's leadership and executing actions with a complete chance with a minimal cost. Exchanges are scratched off the permanent, sealed, solid and durable system.

Blockchain has obvious advantages

Blockchain is a very simple kind of innovation. Every user within the framework has access to all information. In contrast to banks and other organizations that keep their data secure The blockchain provides simplicity and transparency, which customers appreciate.

The information that is stored in the Blockchain is completely secured. The framework is designed to ward off hacking attempts and tries to interfere with the information. In the event that someone wanted to access data that was stored in a blockchain that had 3000 clients. The person would have to hack into each of the 3000 computers in order to change the information. The blockchain's purpose is to thwart any adjustments to data that is entered. If the data has to be changed, it's done via consensus, and the process can be slow because it impacts other entries.

Every exchange through the Blockchain are dealt with in a matter of minutes. Installments are processed and compromises completed in a matter of minutes. External frameworks typically require several days for payments, but the blockchain has only ten minutes and, in some instances the blockchain takes significantly less time.

There aren't any expenses such as charges, exchange or other costs included. They are eliminated when the mediators and brokers are replaced.

Blockchain's limitations

Although the blockchain technology is remarkable in any way but it's not without obstacles. The technology is mostly new, and issues are encountered that were not previously considered.

The blockchain is constantly evolving. It's constantly growing with new coins being added. It is a major problem due to the fact that there are challenges with regards to the quantity of additional space required.

The system isn't yet up to speed in the handling of exchanges. Blockchain can handle 7 exchanges per second, which is a small amount considering VISA is able to handle up to 56,000 exchanges every second.

There is a sense of confidence that these issues will be resolved and this is a given. There are many computer programmers , software engineers and researchers working in labs trying to enhance and improve the blockchain. It is widely accepted that over time, these issues will decrease or even be completely eliminated.

Chapter 5: The Best Ways To Invest On Ethereum

& Cryptocurrencies

Ethereum

A single of the more renowned digital currencies on the market in the present is Ethereum. It is one of the few digital currencies that are attracting serious interest from financial backing organizations around the world. Since the beginning of March 2016, Ethereum has acquired more than 2700% of its value. This is the fastest and most significant advancement of any cryptocurrency ever.

Ethereum continues to gain momentum due to a real issue for financial backing all over the world. The digital currency that is elective, Bitcoin, has been very unstable in recent times. A sharp decline in cost is due to rules imposed by the Chinese government. Controllers from China have imposed restrictions on ICOs, also known as the introductory coin contributions that are financial requests for financial backing organizations to convert funds in virtual currency.

Ethereum isn't like Bitcoin. It's not an actual cryptographic currency but it is a system with lots of potential. Financial experts believe the cost of Ethereum will depend more on the basis of its innovations than the entire cryptographic money market. A majority of experts believe that the price of Ethereum is likely to rise substantially during the final quarter of 2017.

It is a good idea to invest in cryptocurrency

Virtual currencies like Ethereum or Bitcoin are among the most well-known investment options on the market today. These advanced standards of monetary technology could become global monetary standards before long. They may become the preferred installment method for global exchange as well as travel and tourism.

There are many ways of investing resources into these forms. Some financial backers prefer to place their resources into a variety of digital currencies, in addition to the interest they earn in various security schemes. Others believe in that they should purchase and hold major

digital currencies means having an offer to participate in these important initiatives. If an investor puts the assets they have in Bitcoin or Ethereum the value of their assets over time will likely to exceed $10,000.

Buy and keep

One of the most successful ways to venture with digital currency is to purchase and keep. In the event that the value of Ethereum has increased over 2700% in just two years, it's likely to increase dramatically over the next five years. Similar developments are expected to happen with Bitcoin as well as other major cryptocurrency.

Beware: Investors are advised to put aside cash that they be liable to lose. If the venture be unable to pay back their investment the financial backer must at all times be able to save on expenses for living.

The significance of investing in cryptocurrencies

There are good motives for anyone to invest in crypto. One reason is to hedge against the eventual fall or devaluation of the US dollar. There are some who believe that the dollar will be able to decline.

They have many advantages that are not like other currencies. They are, for instance, not controlled by a government or any other organization which means that no one has the power to stop a user's accounts or deny them the opportunity to create one, and so on many other things.

Many investors are enthralled by the technology behind the technology

Cryptocurrencies and the appreciation that virtual currencies are accessible to all people around the globe.

The potential gains from investing in cryptocurrency are extremely appealing and many investors have experienced an incredible return on investments. The trend is likely to continue for a long time , as an ever-growing amount of people around the globe are turning to virtual currency.

Where do I begin?

In order to convert funds into cryptographic money A financial backing company needs to buy their preferred advanced money through an exchange. There are three or four reputable trades available on the Internet which allow anyone to buy Ethereum or Bitcoin or altcoins.

A few examples of well-known stages and trades in digital currencies include GDAX, Coinbase and Bitfinex. Others include LocalBitcoins, CoinSource and SatoshiPoint. The vast majority of web platforms offer cryptographic versions of currency and anyone can create a bank account and purchase the cryptocurrency they want.

Prior to the year 2016, there was only one notable or legitimate digital currency. Financial backers had much choice and weren't able to search through an array of advanced standards for monetary transactions. This is especially true today as in addition to Bitcoin it is also many other digital currencies. These include Ethereum, Litecoin Dash, Ripple and some more. However, many users trust and believe in Bitcoin.

Digital wallets

Before buying any cryptocurrency customers, financial backers and purchasers should secure a reputable wallet. The purpose of a computerized wallet is to store secure digital money forms that are secure and secure from hackers and criminals. It is essential to acquire an encrypted cryptographic wallet prior to making a decision to invest.

For a long time, there was the notion that computers' wallets could hold digital currencies. It's not true. The virtual currency that are purchased remain within the cryptocurrency. What is offered or granted will be the keys. The buyer receives two distinct keys after they purchase advanced forms of money. There is a private key and a public key. Private keys are most significant , and is displayed in a string of characters. This key grants a customer access to their computerized monetary standards that are stored in the blockchain.

Buy cryptocurrency

With a modern bank account an financial backer will be able to remain on their base or exchange of choice and purchase the money they require. There are many considerations to think about when selecting the stage. For instance, how genuine is the stage? How long have they been there and if they can say they are genuine? Each of these are valid questions which should be asked. Fortunately, the trades listed above are generally trustworthy and trustworthy.

Another thing to consider is the cost of the purchase. Many stages and trades will charge buyers in an unanticipated manner. For instance, Coinbase, which is the most well-known stage among Americans is charged a flat cost of 0.5 percentage and 2 additional4 percent per exchange. Stages such as GDAX are considerably more affordable and have charges that are lower. The stage that a buyer chooses will determine their expenses as well as the charges they'll incur.

Digital wallets based on apps

One of the easiest methods of accessing and buying cryptographic money is with portable advanced wallets. These are apps that are able for a more efficient way to purchase crypto currencies.

An authentic example of an app-based wallet is bread-Wallet. The application is available for both Android as well as iOS. It's designed to be user-friendly as well as secure and reliable. A tiny portion of the value of digital currency that is accumulated is often transferred to excavators, so they verify the computerized wallet before adding them to the Blockchain.

The bread-wallet saves the keys from the phone and removes them from the internet. This is a very secure method of saving digital currency. However, in the unlikely possibility that the phone gets lost or the data is compromised the computerized monetary standard may vanish forever. So it is essential to move the data within the wallet, then remove the financial forms. This is a method to protect them.

An important point to note is that some wallets do not work with all currencies. Certain are more explicit such as bread-wallet, which only is compatible with Bitcoin. So, customers must confirm the limits of their wallets prior purchasing one. Certain wallets, such as Jaxx recognize a variety of financial standards, such as Ethereum as well as Bitcoin. Buyers should always affirm this to avoid from disappointments and delays.

The storage of cryptocurrencies by third party

For those who prefer not to keep their own electronic wallets can save their digital currency with external providers such as Coinbase. Coinbase is the most trusted in the cryptographic money stage. They are able to safely store digital currencies for the benefit of their clients. The platform is extremely stable and has raised over $117 million in funding from investors, including Horowitz along with The Tokyo Mitsubishi bank.

Funds held at this point are guaranteed and Coinbase is able to hold both Ethereum as well as Bitcoin. Coinbase can hold both private and public keys for their clients. For security reasons new financial backers are able to choose Coinbase as the place to store their crypto.

Information Point

In putting money into cryptographic currencies There are certain elements that you must pay attention to establish verifiable or failing financial standards. The most reliable digital currencies have an active development group, and the technology employed is simple and they're able to create flexible networks. However, the failure altcoins talk about a muddled set of specialized benefits as well as advanced MLM similar products and continue to claim they're the next Bitcoin and frequently target naive customers.

It is vital to always be on guard fact that there are many new altcoins that are inventing others and cease to exist. A large number of financial backers have lost money and others have made lots of profits based on the strategy they choose when making investments.

The purchase of cryptocurrency directly

It is possible to avoid each one of the costly expenses and costs when buying digital currencies. This is done through direct purchase. Customers can buy altcoins as well as bitcoins directly through dealers. This usually happens in digital currency trades. The trades currently are seen across Europe, Asia and America.

In the US there are BitFinex, BitStamp and Gemini and In Europe the market is Kraken along with Bitcoin.de. In Asia the crypto exchanges comprise OKCoin, BTCChina and BitFlyer.

Purchasing digital currencies

Each of these transactions requires that the customer open the record using their personal information, such as names and email addresses. It is legal due to tax avoidance rules. When the account is opened and funded, it can be accessed through a variety of methods like by credit card or a direct bank transfer. The accounts are subsidised using the currency of the government, either a customary or official one such as the euro or dollar.

As a financial backing institution it is advisable to choose an exchange that is as close as can be expected, should it occur. there is a possibility that local laws apply, so financial backers can recover their funds would be a good idea should the worst to happen. In any event it is advisable to make trades within stable countries like Europe where the general set of laws are based on.

Financial backers and buyers appreciate the lower cost of trades that have some flexibility. The bigger trades may provide many cryptographic forms of money but at high expenses and much faster. Less substantial amounts with better cost can be obtained through trades that are less expensive therefore it is better to determine what financial backers want to acquire their cryptographic forms of money. The persistence of a person pays off and the those who buy and hold are more likely to benefit.

Aside from the usual trades in which venders and purchasers meet, there are other locations known as Altcoin stores which offer all kinds of altcoins. Remember that altcoins are all types of money that are distinct from Bitcoin.

Common altcoin trading platforms include Yunbi, Bittrex, and Bithumb among other. There are other websites that provide information useful to buyers, such as posting of the entire crypto trades that have occurred to date.

What is the ideal moment to buy?

In the world of cryptographic currency there is no set of rules that determines the most optimal time to buy the latest coins. However, experts warn buyers not to purchase when cash price is falling or when it's near the top that is an air pocket. It is best to purchase when the cash and businesses are in a steady state and the price is not too high.

A cryptocurrency-based form of money that trades basically depends on two factors. This is the time where a coin is placed in an air pocket about to explode, and the point at which it will fall as low as it is possible following a crash. The nuances of these focuses can be difficult to determine, so it's ideal for investors seeking a temporary boost buying when the trend is upwards, and sell when it is trending upwards.

If you are putting funds into multiple types of financial instruments, it is best to store them in an exchange. If not , it could be too risky to bring malware to each, upgrade, sync and add and much more. In Europe an excellent platform to buy digital forms of currency is Bitcoin.de. However, when you invest your funds into numerous digital currencies it's not a good idea to let them sit on trades. This is seen as being extremely risky. It is a lot better to meet the challenges and to store them unconnected. This is a risk consumers must take.

Storing digital currencies

It is a great idea to use and store the most advanced standards of monetary management. It is possible to achieve this using the right programming, but without the assistance or confiding in a third party. There is a variety of open and free programs to secure the storage of digital currency. However, all of these data sources must be secured and the correct malware must be used. This is due to the fact that should something occur, like the computer being damaged or crashing, it could trigger a malware attack. There are numerous possibilities to increase the the capacity in digital currencies. If the proper steps are taken to safeguard the digital currencies, they'll be safe far from thieves and hackers.

It is a good idea to invest in Ethereum

Alongside putting funds into the most well-known cryptocurrency, Bitcoin Many financial supporters are now focusing their attention on the second-most well-known one, Ethereum. Ethereum is an innovative blockchain technology which is more of like a convention than a currency. It is, however, thought of as a form of money because it uses tokens referred to as ethers, referred to by the name ETH in the world of digital currencies.

Ethereum its popularity keeps growing across America and around the world. Financial investors are putting their money into this cryptocurrency because of its remarkable performance, its potential for the future and its unique design that is distinct from the likes of Bitcoin.

Ethereum currently has an estimated market capitalization of $27 billion, which is comparable to 40% of Bitcoin. This is a huge feat given the phenomenal growth of Bitcoin as well as the slow or mediocre performance of the other altcoins. In its most efficient form, Ethereum was esteemed at $400. It is currently at $300 at the time of October 2017. Many financial institutions attribute its success in the same way as Bitcoin which has been performing exactly the same way as Bitcoin. This is why it is often referred to as Bitcoin's younger brother. Ethereum is considered to be a precious stones in the realm that is digital currency. This is due to the fact that it is a stone with intrinsic value and fundamental value. Bitcoin is more expensive than gold. It is typically traded for its value, but is of no contemporary or basic value.

Digital wallet

The first step is to get an advanced wallet that can store the cryptocurrency. Some wallets are not able to keep Ethereum and the recommended ones is on Coinbase.

Coinbase is free and offers additional services and an application to Smartphone clients. It can store Ethers, which are the true unit of Ethereum.

Where can you purchase Ethereum (ether)

Investors who want to buy or invest in Ethereum are actually buying Ethers. Ether is the name given to the money used to purchase ethers on the Ethereum stage. There are online financial firms as well as cash-based cash platforms that computerize that allow ether to be purchased. Like we've previously clarified it is essential to ensure that you have a steady supply before buying. For instance, the reliability of sellers should be established.

Coinbase is a trusted platform where various forms of digital money can be purchased. This includes Ethereum, Bitcoin, and Litecoin. The site provides wallets for free for crypto coins, so it's an enormous advantage to use this platform.

Important things to remember

It is crucial for those purchasing to remember that when buying ethers, they are buying cash , not investing resources into an asset. Ethereum is not a form of stock, and there are any benefits, profits or payments. The reason for investing money into Ethereum is done with the hopes that, in the future, other financial backers will be happy to buy identical ethers but at a higher cost.

Coinbase is a great place to buy ethers as well as other forms of cryptographic money. In the case of example, when the buyer opens a transaction on Coinbase it is possible to borrow money and then buy the ethers through the website and even keep them there. The platform provides an advanced wallet that is customers for no cost and users can even earn cash back when they buy $100 worth of digital currency.

It is advisable to invest to a minimal amount and gain little purchase of Ethereum over time. Financial buyers or backers are advised not to make huge purchases in light of the possibility that a drop in the price of Ethereum could create misfortunes. In addition, financial backers must be careful when investing money in which could be lost should the funds fall.

It is also possible to create the record to the degree that the Ethereum purchases are consistently made. This works well, so that, each time the ether worth $100, $50 or $1000 are purchased in accordance with

settings and settings and. The most important thing to remember is that the cost of Ethereum is subject to change from month to month. This way, distinct Ethereum amounts will be acquired at any time when bought with the same sum of money.

Strategy of Buy and Hold

If a financial backing company purchases Ethereum most likely the best strategy is to keep it for a time period at least one year. It is not uncommon to see massive value variations within this time. The goal is that buyers earn a significant profit from their investment.

In the past two years, Ethereum has acquired more than 2700% in value between January 2016 and October 2017. The idea of securing it over a reasonable period of time is likely to be rewarded with positive yields. Tolerance is crucial here. The long-term investing in cryptocurrency and in particular products like Ethereum is the most effective method to build wealth and hedge in a positive way.

Avoid day trading at all cost

Investors may be enticed by the opportunity to join in day trading requests to make quick gains. Day exchanging is of buying a few items typically during the daytime when costs are lower, and selling the item as the price increases. It is a highly theorized exchange that is extremely dangerous. As a rule the financial backers will lose money. Even the most experienced traders have difficulties engaging in day trading. The most effective method of investing using Ethereum is to buy and keep. It is also beneficial to buy small amounts and build your an enterprise regularly.

Sometimes, buyers look to gain from the smallest shift in cost. In reality, many financial backers are persuaded to offer their services at the lowest increase in price. This isn't the best practice due to the fact that a lot of other companies will increase their prices and follow suit, resulting in the value of their investments to plummet and instability within the market.

It is always the duty of the buyer to safeguard their Ethereum. The benefit of keeping digital currencies on Coinbase can be that it is totally safe as the site is secured and all financial requirements are covered.

Chapter 6: Markets Are Altered By

Cryptocurrencies

In the past couple of months and over a long period of time, several countries around the world have authorized digital currencies , and properly provided them with real installment plans. Other countries like Poland, Germany, Switzerland and Australia have passed laws that allows the use of cryptocurrency.

In the future, exchanges in the business and financial sectors might accept installments in cryptographic types of money , like ethers or bitcoins. It also implies that merchants, stock trades dealers, venture banks, and dealers might embrace blockchain technology. This would make exchanges more open and safe while also making installments easier and reducing the costs of major businesses.

Companies that deal in cryptocurrency are now raising funds by using methods like ICOs. An ICO is basically an underlying coin offering is a method of raising funds by offering digital currencies to financial backing organizations at a low cost. This is a more efficient method to raise funds in spite of the fact that it's similar to IPOs. Financial backers will have an additional device to include in their portfolios.

Many financial institutions are now adding cryptographic currencies to their plethora of speculations. Digital currencies are often viewed as wares, which could add to the portfolio of financial backers and increase the value. Common cryptographic currencies such as Ethereum Bitcoin, Bitcoin, and even Litecoin, are being exchanged on the internet in a way that is efficient and effective.

The market for cryptographic currency is currently worth more than $145 billion. This is a huge market that cannot be ignored or overlooked. It is changing the structure and perception of the monetary and financial sectors. This the year (2017) the value of one Bitcoin beat gold by a significant margin. The market's growth has grown rapidly and is significantly more than most people think of.

Blockchain

The blockchain will definitely affect the way that businesses operate. Consider the strength and stability of this technology. It is the foundation for all digital money forms which ensures that all exchanges are transparent, verified as well as unreliable and long-lasting. Each of these is features that are attractive in the world of money.

As of today, there are several organizations trying to develop applications based on blockchain technology for use in the area of money. Financial exchanges, for instance are able to use blockchain technology to ensure that transactions are secure and dependable, and also to protect the brokers who are a bit shrewd from being harmed. It could also provide new levels of honesty and openness to customers would like.

The likelihood is extremely high that blockchain innovations will soon replace existing financial frameworks. On the other hand they will take on the same role as traditional shops and online stores or print media or computers.

The current frameworks used in banks, as well as by other financial frameworks are complex and risky, exposing customers and institutions to unnecessary risks and costly operational expenses. Blockchains, such as the Ethereum-based ones use smart contracts. They are viewed by some as untrustworthy however, many eyewitnesses appreciate the fact that they are highly secure and robust against attacks, even though they are not affected or influenced by government policies.

Cost savings

They have certainly decreased the cost of exchanges. One method they have used to achieve this is by removing mediators and delegates. A majority of middle people seek out the cost of expenses as well as slow cycle. Once they die, the process becomes less costly and more efficient.

The blockchain also eliminates the need for administrative centres in which a large number of individuals work. If the administrative center is able to be shut down, businesses can set aside money as well as exchanges. This will make them more affordable and, consequently, more affordable. The speed of processing exchanges will also improve.

It's a known fact that cryptographic currencies and the blockchain are undergoing a transformation.

the scene of the money and monetary areas. They've presented more efficient methods for execution and improved and more reliable platforms that can often reduce the cost of operations, reduce schedules and help increase efficiency. It is commonplace that as the years and duration pass there will be an increasing amount of plans will be adopted that make use of both blockchains and cryptocurrency.

Chapter 7: What Is The Relationship To

Cryptocurrency As Well As Blockchain?

The attention of the media on blockchain technology and cryptocurrency has increased in recent times. Everybody has heard of these two concepts however, few people are aware of the relationship and the distinction of Blockchain as well as Cryptocurrency. Blockchain technologies are being put in both established companies and new ventures. But, it could be unwise to get familiar with the terms and not understand the relationship between them prior to investing in digital currencies.

The words Blockchain as well as Cryptocurrency are often used in conjunction. Though they are two different techniques, the two are connected. To understand the connection between them you need to understand each word in its own way.

What precisely is Blockchain?

Blockchain is a database of all cryptocurrency transactions that are conducted across the globe at any time. It's a system which stores the data in an encrypted format, which makes hacking or tricking the system very difficult. Blockchain is a database that is public which shares all the information regarding cryptocurrency transactions across all the networks of distributed systems.

Instead of keeping information on a central server, the data is spread across hundreds or even millions of computers around the world every computer is granted the ability to access it. Everyone has access to the data, which includes trading and mining cryptocurrency. This data does not have to be subjected to central authority or oversight. Transactions are classified in "blocks," which serve as storage devices for data. Blockchains are distributed database that arranges transactions chronologically. It is composed of the strings that have been successively created of each block that have been created.

Is cryptocurrency a real thing?

It is store of value which can be used to buy or sell goods, services real estate, products, or other precious commodities. It is the case that Bitcoin as well as Litecoin have become two of the most well-known cryptocurrency. They are both cryptographically protected against flimsy transactions and are usually not issued or regulated by any central authority. Instead, the users are responsible for the governance of the network. The term "cryptocurrency" is often used to refer to digital coins or tokens. To learn more about cryptocurrency, go to the-crypto-superstar.com/de.

What is their relationship?

Blockchain technology isn't an option to use to use cryptocurrency but an essential element of cryptocurrency. Additionally, Blockchain and Cryptocurrency have similar origins. They're not at the same level. But, when confronted with each other, Blockchain triumphs. Blockchain isn't just used in the financial sector. It offers a variety of options that could change the market over the coming years.

Since all Bitcoins were stored in the database at the time that Blockchain was created as an online database that the names are that they are interchangeable. The year 2009 was the first time

Blockchain did not get recognized as the original name. Because transactions were separated into blocks and connected with a mathematical process which creates a hashcode Blockchain was named that.

Blockchain and Cryptocurrency Future Cryptocurrency and Blockchain Future

In 2022, Blockchain is predicted to generate $11.7 billion worldwide. As established organizations and new startups are able to profit from the power of Blockchain technology, it is becoming apparent that the issues brought about through Blockchain and Cryptocurrency go beyond the realm of economics.

It is tempting to become distracted by the hype however the pace of technological advancement isn't slowing down. Despite the hype surrounding that bubble, a lot of people are still convinced that blockchain-based currencies can be an investment that can be held for the long term.

Cryptocurrency and blockchain technology have facilitated the development of many industries. As per MIT Technology Review, cryptocurrencies have dropped by more than 90% since the year

2017 however, they believe that blockchain technology remains essential to innovation across all industries. The constant acceptance of this technology indicates that it is our future.

Bitcoin and Blockchain Relationships Bitcoin Relationship

We all know that Bitcoin is a form of digital money However, Blockchain technology is what makes it possible. Blockchain is the tech that lets bitcoin be created. Bitcoin is useless without blockchain technology, since it is impossible to make secure transactions using it. Blockchain is a database that has been verified which guarantees the authenticity of every transaction. This means that you are safe from fraud and double-spending.

Understanding the connection to bitcoin as well as Blockchain is vital. Distributed systems can track and verify transactions with Blockchain technology prior to adding a new block on the chain. This provides greater security and protection. The history of the chain cannot be modified with any type of computer. A consensus-based system will eliminate the dependence on central banks and banks to maintain accurate information.

The current Blockchain debate is limited to its use within the financial sector. In the near future, it could be used for document-keeping property deeds, property deeds, as well as contracts. A lot of hosts consider Blockchain as a complicated technology.

What is Blockchain and how does it work for Crypto Data Storage?

Blockchains maintain an entire chain that connects all the blocks' data and the blocks. As new data is added to in the chain, it is created the block. Once the block is completed and connected to the block that preceded it. This allows records to be classified chronologically. Each block contains information on different transactions.

Let's examine an example of an Excel spreadsheet. The data is also contained within the worksheet. It is intended for only the handful or even a few of users to access or edit the data. A vast database holds large quantities of information that could be accessed by many users at once. They are usually managed and owned by one single entity or the individual. The Blockchain however, on the other however, is not centralized and is not owned by any one individual or company. Therefore, the Blockchain is reliable and secure.

The connection between them is indisputable. Blockchain technology is an important element of cryptocurrency. Blockchain is able to be a banking system or, more precisely it's distributed ledger systems. Because of its digital advantages and capabilities, it can also permit users to create blockchains using only a single parameter.

Ethereum

Ethereum isn't a bad relative to bitcoin. Ethereum is not a sluggish cousin to bitcoin. Ethereum blockchain is at the heart of various crypto-verse trends, such as DeFi, the metaverse and NFT obsession. Here's everything you should learn about investing in the Ethereum, which is the currency of the blockchain , and the second largest cryptocurrency in the world.

The momentum generated by bitcoin the leader in cryptocurrency is a catalyst for other major digital assets, like Ethereum (or Ethereum) and ether to new levels. In fact, Ethereum outperforms bitcoin by an enormous margin, and is stealing some of bitcoin's thunder. In 2021, Ethereum's worth was up by over 400%, which is more than bitcoin's modest gain of 62 percent (as as of December. 31,).

Although it's still helping investors who are dissuaded by the high cost of bitcoin, Ethereum is much more than just a "cheaper bitcoin." In recent times, Ethereum has made rapid technological advancements and is gaining wide acceptance among developers, as well as increasing the interest of investors.

Does Ethereum's rapid growth continue into the new year? Let's look more deeply at its capabilities and features and the elements that can determine its future growth and the best way to purchase it.

What is Ethereum?

Ethereum is a blockchain which is a non-profit software system that is open source that is used for creating decentralized applications. Ether (ticker Ether) is the platform's primary currency is used to pay transaction fees and algorithms that are offered on Ethereum. Ethereum blockchain.

Ethereum as bitcoin is built upon blockchain technologies. It is, however, Ethereum has earned a advantage in the market by allowing individuals to build applications that utilize its blockchain technology, including smart contract capabilities. On the blockchain smart contracts execute themselves agreements. They can be used to verify and record transactions between sellers and buyers without the need for an intermediary or central authority.

Ether is a great source of potential in the real world due to the fact that it is the Ethereum blockchain can be described as an sought-after technology for the development of smart contracts, which highlights its value not just as an application platform for developers but as a global financial service.

Anyone who wishes to use an intelligent contract should be able to pay for the cost with Ethereum. There is also an increase in the number of new uses cases that are that are built on Ethereum. Ethereum blockchain, as per Michael Zagari, an investment advisor at Mandeville Private Client Inc. This includes Decentralized Finance (DeFi) web 3.0 Non-fungible Tokens (NFTs) advertising, gaming identity management, as well as the management of supply chains.

These usage examples illustrate the importance of Ethereum DeFi, an application that allows users to borrow, lend or trade bitcoins on an uncentralized blockchain network, without the involvement of brokers or banks. NFTs, which is a $10 billion business is the fastest increasing Ethereum invention.

Ether is also becoming more accepted as a payment option as more businesses are accepting the currency for purchases and services. "We're nearing seeing the worldwide shift towards DeFi speed up a few notches," Green adds, noting that the Ethereum blockchain has now become an essential element of transactions that don't require traditional banks.

Is ether a wise long-term investment to invest in 2022?

Ethereum is in its beginning stages and the potential for growing acceptance by the masses is thrilling. The continuing growth of Ethereum has opened the way to a brand new array of features, which is which gives ether a new lease of life. This is great news for the potential of blockchain technology as well as the price of Ethereum. Ether is a good investment for the present and the future due to the superior technology it is built on. Its price will probably outstrip bitcoin in the next five years.

Recent improvements in the acceptability of ether price, as well as media interest have helped draw large-scale investors adds this is an additional reason for the cryptocurrency's increased visibility and demand.

In their constant pursuit of their pulse on the cryptocurrency markets, Ethereum supporters recently deployed the Eth2 update to improve the structure of the network. The value as well as the popularity Ethereum are increasing as investors look forward to the revolutionary transition to Eth2 which will make the network more efficient, cost - efficient and safe.

Businesses are examining and integrating Ethereum to their existing processes since it is the most popular cryptocurrency. With that kind of brand recognition and market dominance compared to other networks that are decentralized, and also its smart contract capabilities Ethereum could be thought to have an edge over competitors.

The early adopters of this technology reported positive returns on their investment. Although no one can assure that it will be similarly in 2022. I think Ethereum appears to be on the right path to the widespread acceptance. If you are thinking about investing over a long period of time, Ethereum might be an ideal long-term investment.

Many crypto experts agree that the there is a high demand Ethereum will grow in the future, due to a variety of factors, including its status as the most secure blockchain as well as its ability to scale and the ability to complete transactions faster, but with a lower price.

The Metaverse's Ascension

The emergence of crypto-related phenomena like DeFi NFTs, DeFi as well as the metaverse, have increased activities on Ethereum blockchain, driving up demand for Ethereum. On October 1, Facebook changed its name to Meta in order to show its growing importance on the virtual realm where people can work as well as play. Metaverse's market is projected to increase from US$500 billion by 2021 and reach US$800 billion by 2024.

The demand for Ethereum-powered metaverse currencies is predicted to grow in sync. Decentraland, The Sandbox, and Axie Infinity are examples of virtual reality platforms which are based on Ethereum. Ethereum network.

In the event that all these applications meet different needs across various industries and are developed by leveraging the Ethereum network, one can be able to conclude that demand for the Ethereum, the native cryptocurrency, is likely to rise and, consequently, have a greater value.

NFTs are frequently viewed as the most innovative Ethereum invention. The digital tokens are affixed to items like music, art as well as other media, with blockchain technology. This helps in proving the ownership of and the validity. Since that the vast majority of NFTs come from the

Ethereum blockchain as well, the NFT phenomenon has led to an ether bubble which shows no sign of slowing. Additionally, the ongoing equities market volatility has led many investors to invest in ether as a possible alternative that could provide non-correlated returns compared to traditional assets. Ether could offer a desirable return-on-risk if you purchase it and keep it for a long period and spread its high risk of volatility over a longer time.

What are Smart Contracts?

Smart contracts are a part that are part of Ethereum software that work using the Ethereum Virtual Machine (EVM). It is an uncentralized "world computer" that makes use of the computing capacity of every Ethereum nodes. Any node that has computing power is compensated with Ether tokens.

Smart contracts are referred to as smart contracts since they let you create "contracts" which can be executed automatically when specific conditions are fulfilled.

You could think about creating a crowdfunding site that is based on Ethereum similar to

Kickstarter. A person could develop the Ethereum smart contract, which mixes funds and transfers them to someone else. This smart contract can declare that once $100,000 worth of currency is added to the pool and distributed directly to the person receiving it in full. In the event that the $100,000 threshold isn't completed within a month all money will be returned to its original holders. Of obviously, Ether tokens would be utilized instead of US dollars.

All of this would happen after the smart contract code that executes transactions on its own, without the requirement for an intermediary to keep the money and give their approval for the transactions. For instance, Kickstarter charges a 5 percent fee in addition to the 3%-5 percentage processing fee and a total charge of between $8000 and $100,000 for an investment of $100,000. Smart contracts do not require paying fees to a third party like Kickstarter.

Smart contracts are utilized to fulfill a wide range of needs. Like the way software libraries function developers can create smart contracts that aid in supplying functions that are not available in different smart contract. Smart contracts can also be utilized as a basic application for storing information using Ethereum blockchain. Ethereum blockchain.

One must provide enough Ether to pay a transaction fee to run smart contract software. The amount will depend on the computing power required. This pays Ethereum nodes in exchange for their participation as well as computing power.

Smart Contracts are utilized by CryptoKitties

CryptoKitties is a game that claims to be to be "one of the first games that is built using blockchain technology" is among the most popular apps built with smart contracts that run on Ethereum. Ethereum network.

CryptoKitties are basically a type or digital "collectible" that is stored in the Ethereum blockchain. CryptoKitties is a perfect example of Ethereum network's capability to store and exchange digital items.

"Breeding" leads to an increase in the number of CryptoKitties. This involves selecting two of the most basic CryptoKitties and then using Ether tokens to control an intelligent contract. The contracts will create a brand new CryptoKitty based on the two cats selected. The kittens, as details about the breeding process are kept in the Ethereum blockchain's public ledger.

CryptoKitties, which are held within the Ethereum blockchain ledger, can become "owned." You may trade or sell them to others, or you could buy them. It's not like using apps on smartphones to purchase or trade cats. The pets are usually stored on the servers of the app and you could lose your precious digital pets in the event that the company decides to shut down the app or blocks your account. But, as CryptoKitties are stored in the Blockchain, that isn't feasible. There is no way to keep your pets away from you.

In December 2017--coincidentally, around Bitcoin's all-time high prices--people spent the Ether equivalent of more than $12 million for CryptoKitties, with the most expensive CryptoKitty selling for over $120,000.

How Is a Wallet? This is Everything You Need to Learn

Digital wallets are now an integral part of our daily life. What exactly is a blockchain-based wallet and why are you interested? Let's get started. A blockchain wallet is a digital wallet that allows users manage and store their cryptocurrency assets, such as Ethereum, Bitcoin, Ripple and many more. This can facilitate cryptocurrency and local payments. It's free to set up an e-wallet with Blockchain Wallet, and you can do it on the internet. Customers must supply an email address and password to access their accounts. The system will send you an email asking for confirmation of your account.

The user will need an identification number for the wallet once the wallet is created. This is a unique identification. Users of wallets can login to the e-wallet either by logging into the Blockchain website or downloading it using the mobile phone. Blockchain Wallet's Blockchain Wallet interface displays the balance of the wallet's cryptocurrency and the most recent transactions made by the user. Prices tables also are accessible to users. Check the value of the fund's assets in the currency of the user. There's also an educational section in which you can get to know more about cryptography as well as get the most recent news.

The Blockchain Operation Mechanism of the Wallet. Operation Mechanism

Users can request a specific amount of Bitcoin or any other cryptocurrency from a different party. The system creates an unique address that can be given to a third-party or transformed into the form of a QR code. QR codes, just like barcodes, can hold financial data and can be digitally read.

When a user sends an application, he/ creates a unique email address. When someone offers them the information they need, they may also transfer cryptocurrency. The method of sending and receiving is similar to PayPal but bitcoin is the preferred method. PayPal is an online payment service acts as intermediary between customers and banks. Additionally, it facilitates credit cards, it encourages the transfer of funds via financial institutions.

Users can also exchange Bitcoin for other crypto-assets , and in reverse. It is a straightforward technique that does not compromise its security. blockchain wallet for crypto. The user may be able to see the information that shows the amount of money they'll receive with an exchange rate currently in place. The amount of money charged varies based on how long it takes the client to complete the transaction. It should take about a couple of hours, based on the number of transactions transferred to each currency's blockchain. Users are advised to contact the support number if the process is taking longer than six hours.

There are only six cryptocurrency that can be traded in a wallet on blockchain: Bitcoin, Bitcoin Cash, Ethereum, Stellar, Tether, USD Digital, Wrapped-DGLD and Lumens. Users can also buy and sell cryptocurrencies through the wallet's Buy Crypto interface. Marketing and purchasing services aren't accessible all over the world. Customers can transfer funds through their bank, make use of credit card or utilize their cash balance to fund the purchase. The maximum and minimum purchase orders amount is $25,000 and $5, respectively respectively.

Charges for Wallets

It is worthwhile to mention that the business uses an approach known as dynamic payments. This means that the commission payable for each transaction can be contingent upon a number of factors. The value of the transaction and the conditions of the network at the moment of the transaction can have a major influence on the price. Many transactions are handled by the powerful computers referred to as miners. Miners usually handle the highest commissions on transactions, and it's profitable financially for them.

To ensure that transactions are processed in time to ensure that transactions are completed on time, the service offers a priority charge. The processing time is around one hour in the average. However, certain recurring payment options are more affordable. It can, however, take a little more time. The customer can also alter the cost. If the user offers a lower price but the transferor's transaction could be denied or delayed.

Consumers need to be aware of the security of their wallets. The users could lose control of their funds in the event that they lose their wallets.

There are a variety of levels of security within Blockchain Wallet. Blockchain Wallet. It protects the cash of the user from potential hackers and even the business itself.

Security

To protect customers, Blockchain Wallet profiles, as with other digital services, require passwords. Blockchain, as a Blockchain company however it does not store user passwords. Therefore, if the password gets lost, it can't be reset. This protects against insiders using the company's cryptocurrency. If a user forgets their password, their account will be reset using an mnemonic sequence.

Mnemonic seeds are used as an alternative to a password. If a person is unable to access their smart phones or other device they can used to locate wallets that contain any cryptocurrency. Mnemonic seeds of customers are not stored in the database of Blockchain. Blockchain firm. Since these seeds conform to industry standards, wallets can be saved even if the company is shut down.

Different security options protect the wallets of users from external threats. Blockchain Wallet also enables users to use two-factor authentication to reduce the chance of phishing attacks, and IP allowlists that prevent logins from devices that are not familiar with the blockchain. It is also possible to limit access to Tor. Tor Network, which stops hackers from hiding your IP address.

Ethereum 2.0

The next major version of the Ethereum cryptocurrency network, known as "ETH 2.0," claims to address many of the network's important issues, which range from high GPU charges to environmental damages. Let's take a look at the proposed changes and what they may mean in the near future for cryptocurrency.

What is Ethereum 2.0 and when will it be released?

Ethereum 2.0 can be a commonly employed term, which generally is used to describe Ethereum's anticipated transition from proof-ofwork to proof-ofstake which will eliminate Ethereum mining. From January 24, 2022 it is expected that the Ethereum Foundation no more refers to this version as "Eth2" or "Ethereum 2.0." The foundation instead refers it as "the merger" as well as "the docking."

As we'll discover later, as we'll discover later on, the Ethereum system's dependence upon computational resources in order to create an agreement ("proof of works") has led to the soaring GPU prices and environmentalists' disapproval. Due to the wide-spread usage of NFTs and the fact that many utilize Ethereum smart contracts that validate tokens tied to artworks These challenges have become more important. The transition to proof of stake which removes the requirement of GPU mining is expected to solve the issues mentioned above.

For a long time the foundation has stated they will be able to transition Ethereum 2.0 is scheduled to occur in the second quarter of 2022.

A Quick Overview of Ethereum 1.0

If you're not familiar with Ethereum consider it a huge virtual machine that is powered by the internet. Similar concepts can be observed if you've played with an emulator old MS-DOS games, or used to operate Windows on the Mac. A computer that was programmable was running in program software (rather rather than hardware) in conjunction with another platform in both cases.

A virtual machine is not a machine running on one computer, Ethereum is an open virtual machine made up of a multitude of computer (referred by the name of nodes) connected by the blockchain. Nodes are able to launch "smart contracts" software programs running on an Ethereum virtual machine. Since Ethereum is a distributed and flexible network the size of the virtual machine can change at any time when nodes join or leave the network.

Payments of Ether (a currency that functions within the Ethereum network's applications) encourages people hosting these servers and provide to the power of computing (known by the term "mining") in order to run smart contracts and verify the order of transactions that are recorded on the Ethereum blockchain. This process of verification is referred to informally as "consensus."

Today's Ethereum issues

To comprehend why Ethereum must be upgraded It is important to first understand its flaws. Ethereum's engineers and experts have identified a number of major problems with the way Ethereum operates, and they consider these issues to be hindering the widespread acceptance and use Ethereum applications. Here are some of the most important concerns:

High Gas Fees "Gas" is the main reason for keeping the Ethereum network operating. It is a fee that is paid to miners who provide the power needed to process transactions that allow the network to operate. The price of gas is a market value that fluctuates based on the amount of resources needed in the Ethereum network. The higher the price of petrol higher the demand. The greater the amount of fuel someone is willing to pay for, the faster the transaction can be completed. If the demand for Ethereum applications increases the price of gas could be too expensive, usually making it more expensive to complete an exchange than the value of the currency being exchanged. For instance, in some cases buying a cheap NFT could result in more gas costs than the value of the NFT itself.

Power Consumption: At the moment, reaching an consensus regarding the Ethereum blockchain is dependent on cryptographic challenges that have to be resolved with the help of Ethereum networks nodes which is referred to by the term "proof that work." The more well-known Ethereum grows, the more computing power is needed to verify its blockchain, which causes nodes on the network to consume more energy. This is why there's been a lot of criticism that the Ethereum network can cause pollution that damages our ecosystem.

Disk Space Consumption As the Ethereum network grows in size, the operation of an individual node becomes more difficult due to the fact that it is evident that the Ethereum blockchain history takes up larger disk spaces. This limits who can operate an entire node (by increasing the cost of running one) which limits how many nodes can be placed available on the network.

Network Congestion: During periods that are characterized by high demand for computational resources Inefficiencies in the way Ethereum functions can lead to problems in the communication between nodes, which delays the execution of smart contracts. Due to this congestion the level of sophistication of apps that can be able to operate within Ethereum's Ethereum network is restricted.

GPU Prices Ethereum's consensus system (dubbed "Ethash") is specifically designed to be lucrative to mine using graphic cards for consumers. The more demand there is for computing through the Ethereum network and the higher the amount miners will be compensated (in fuel costs) and they will purchase more GPUs to earn more. This means that there is a dearth of GPUs which could cause the cost for graphics card to rise. The high GPU price has a significant impact on other GPU uses , such as game play and neural network.

The Planned Solutions

Ethereum "merge" artwork

Since the time of its creation in 2013 The Ethereum Foundation along with its founder Vitalik Buterin were aware of the problems mentioned previously (and began to address them in the year 2015 when Ethereum was launched.) But as the popularity of the Ethereum Foundation grew it became increasingly difficult to make improvements and updates. Any changes to the network require the consent of at the minimum of 51 % of Ethereum nodes (if the majority of nodes don't agree, the network splits or splits into multiple networks). Here's a brief overview of the details of what "the merger" and other enhancements will be able to do to fix the issues.

Moving to Proof-ofStake

After "the merging," Ethereum will no more be able to create consensus via proof-of-work. This process requires energy and processing power from miners. Instead, it will employ a proof-ofstake mechanism that requires validator nodes to take on (or "stake") an certain amount of Ether money to validate block transactions that are on Ethereum's blockchain. Ethereum blockchain.

To add blocks, validaters are selected randomly (confirming transactions as well as operating smart contracts.) If they disconnect at the beginning of the transaction or provide incorrect numbers, they could lose some or all part of the staked Ether. This risk provides an incentive to follow the right path and validaters will receive compensation with Ether to compensate them for the effort they put into it.

In the case of proof of stake validators will have to work hard to construct blocks for the Ethereum blockchain however not nearly as much as they would when dealing with cryptographic issues. In the end, the proof of stake system is expected to drastically reduce the Ethereum energy use of the network as well as reduce barriers to entry (you will not require a costly massive GPU to earn bitcoin as an authenticator). This could also lead to increasing the number of nodes in the network as it's simple to join a pool of nodes. The more nodes, the more computing power and less centralization which increases security of the network.

The switch from proof-ofstake to proof-of-work will likely lower the demand for GPUs however they could continue to be used to mine crypto , as miners that previously used to mine Ether change their mining equipment and methods to other crypto currencies. Graphics cards' costs could fall slightly in the event that GPU demand decreases, but there are other factors to consider in the current graphic card shortage.

The process of switching Ethereum for proof-of-stake is a complex process that started with creation of Beacon Chain, a form of a parallel consensus layer that was centered around staking Ether that eventually merged with the Ethereum network. Ethereum network. This is why the term "the merging" was created.

It's the Spread of Sharding

After "the merger," Ethereum's developers want to roll out a new major update called "sharding," which divides the principal Ethereum blockchain in smaller pieces, referred to under the term "shards."

The entire Ethereum blockchain's history is now four gigabytes. Full nodes aren't required to host the entire amount. But under the new layout, blockchain activity will now be split into 64 pieces, meaning that each node only needs to hold 1/64th of the Ethereum blockchain's size as it is today.

Through reducing the the hardware requirement, sharding can be expected to lower the entry barriers for operating a single node. This could create additional nodes permitting the network to expand in terms of capacity. Sharding can also boost the amount of transactions the Ethereum network is able to handle by spreading the burden over multiple nodes, which could reduce the cost of gas.

Sharding is expected to be available to the Ethereum network in 2023. No exact date has been announced.

Can Ethereum 2.0 decrease the cost of gasoline?

Since "Ethereum 2.0" is currently a reference to various things and is separated into various goals which will be realized in the future, it's difficult to determine with certainty if it will reduce gas prices.

It is believed that the Ethereum community remains skeptical of the switch from evidence of stake ("the merging") will result in lower gas prices and the Ethereum foundation is not able to warrant the outcome. Gas prices are determined by demand and every Ethereum block is limited to a certain amount of computing space. Sharding can lower costs by increasing the processing capacity that is available on Ethereum's Ethereum network, but it's not expected to occur until 2023 for the primary Ethereum chain.

However, some experts suggest that the decline in Ethereum gas costs could necessitate the creation of "Layer 2" programs that are built on the Ethereum network. These programs will perform some calculations on its own but rely on Ethereum for the fundamental acceptance and verification.

To summarize, the problem of Ethereum upgrades and their impacts is complex, and it is dependent on a set of connected conditions--including the size of the network, the value of Ether, the demand for NFTs, and the attitude of the node operators--that may vary dramatically from day to day. It will be interesting to see what happens and what effect Ethereum's upgrades could have on overall cryptocurrency ecosystem.

If we could predict that, the conversion of Ethereum in to proof-of-stake is typically expected to be a game changer. If other cryptocurrency follows this trend, it could possibly eliminate obstacles that prevent certain government agencies or corporations from fully accepting cryptocurrency. In the end, their popularity could skyrocket which would create an extremely crypto-friendly world.

Chapter 8: A Short Overview Of The History Of

Ethereum

The year 2013 was when a computer programmer called Vitalik Buterin who was 19 in the year 2013, wrote his whitepaper, in which he suggested a blockchain system that was flexible and could be used for virtually any kind of transactions.

2014. Vitalik Buterin and Gavin Wood along with their co-founders, have teamed up to have raised funds to develop the Ethereum protocol by selling pre-launch tokens at the sum worth $18million.

2015

The first public release to the general public of Ethereum blockchain was released in July. Additionally, the possibility for smart contract technology is now beginning to be implemented to the Ethereum blockchain.

In 2016, there was a theft of about $50 million from DAO from hackers (short for Decentralized Autonomous Organization). Due to this, the DAO community decides to alter the protocol to recover stolen assets. This results in a hard break and the formation of two distinct blockchain branches: Ethereum and Ethereum Classic.

In 2017 in 2017, the ERC-20 standard was created that outlines a procedure to create assets that are based on the Ethereum blockchain. First Ethereum application to reach the mass appeal was CryptoKitties. The launch of the MakerDAO. Also, DAI, the world's first stablecoin that is based on ETH is introduced by Maker.

The year 2018 is a turning point in the field of decentralized finance, with the launch of the decentralized trading platform Uniswap and the loan protocol Compound. The launch of the USDC stablecoin, made possible by the CENTRE Consortium and is a collaboration between Coinbase and Circle which will result in the issue 1 billion USD worth of currency during its first year.

2020

Transition to Ethereum 2.0 will start in December. It is expected that it could take about two years to make the transition between Ethereum 1.0 and Ethereum 2.0. Although Proof of Work is still being used as a method of achieving consensus in Ethereum 1.0 and 1.0, the first stage for Ethereum 2.0 will see the implementation in Proof of Stake.

2021

There are several updates available that include The Berlin, The London The London, The London, and The Altair. The Berlin upgrade reduced the price of gas and increased the variety of transaction types which were accepted. The London upgrade led to a significant increase in the market for transaction fees , and led to a reduction in refunds for gas. Beacon chain was upgraded to Altair. Beacon chain upgraded its service into Altair in the course of their regular maintenance.

"The "merge" between Ethereum gives those who are bullish of the market for cryptocurrency something new to cheer about.

It is now official that the Merge of Ethereum (ETH) is finally been announced!

MERGE Complete.

What do you mean by "The The Merge"?

The Merge upgrade is Ethereum's eagerly anticipated change from its current "Proof-of-Work" system of consensus to an "Proof-of-Stake" method. This upgrade has been in development for a while.

The Merge is accomplished technically through the two-stage process, called The Bellatrix as well as the Paris enhancements and Paris enhancements, respectively. Bellatrix was the first person to officially initiate the Merge in the 6th of September, 2022 when the time was 11:34.47 Universal Time (UTC). It is believed that the Bellatrix protocol will be an enhancement to Ethereum's consensus layer. Paris is the successor to Bellatrix will function in the role of execution layer as Ethereum switches from proof of-work to proof of stake. This change will take place following the release of Bellatrix. The exact date of Paris is unclear as it is extremely dependent upon the hash rate that is achieved through the proof-of-work method. Paris will be enabled at the time that the Total Difficulty is at a certain point that is known by the term TTD or Terminal total Difficulty (TTD). The threshold for total difficulty also referred to as TTD is a prerequisite that must be met prior to the final block in Ethereum is mined. In other words, TTD is shorthand for the specified amount of hashes that have to be mined before Proof of Stake can be considered to be fully functional.

What follows immediately The Merge is referred to as.

When all is done and everything is in place, it is expected that the Beacon Chain, which is currently operational and will be able to take over the task of checking new transactions with Proof-ofStake and Ethereum's standard Proof-of Work architecture will cease for good. In the Beacon Chain Validators have invested nearly 13 million Ethereum as of the present moment. When the main net which is the main network for Ethereum, Ethereum blockchain, joins along with Beacon Chain, the whole of the history of transactions on Ethereum will be combined simultaneously. This includes each transaction, smart contract and balance as of July 2015 onwards.

Why should we be concerned regarding The Merge?

The Merge is in development for six years and due to the potential ramifications both philosophical and material and philosophical, it is widely recognized as a pivotal moment in the history of the cryptocurrency's growth. After days of market turmoil due to a variety of factors, including to rising rate of interest, this significant event could help boost confidence in the market and bring some necessary optimism. This is after a period in which the market been turbulent. The Ethereum implementation Merge in the words of an analysts, "will illustrate that a distributed and uncontrolled network can operate efficiently." A such a cryptographic fusion is not a common event, and may never happen again.

What effects could The Merge have on me and my cryptocurrency?

In a nutshell it's not all that. From the viewpoint of the user it is clear that the Merge is designed to be smooth and is expected to be. In a step to protect themselves, Coinbase will temporarily stop certain withdrawals and deposits. This will impact new Ethereum (ETH) ERC-20 Polygon (MATIC), The Optimism (OP), PWETH, PUSDC as well as PMATIC and PWETH tokens. However, we don't expect any other networks and currencies being affected. The people that have staked Ethereum However, those who have staked their ETH it is not going to be able to get their balances unlocked , and their ETH won't be capable of being transferred or traded immediately following The Merge.

When the Ethereum protocol is upgraded to its ultimate status, it's predicted that the staked ETH is expected to be released as well as released. According to the most up-to-date forecasts, the upgrade is expected to be completed by the beginning period of 2023. Be on the lookout!

EXPECTED DATE OF MERGE

15 Sep 2022

The Ethereum project has merged its proof of work with visualizing proofs of stake.

You've got ETH Merge questions. We've got the answers.

The race is finished; we've completed the race. The Merge has finally arrived after being repeatedly delayed due to the technical challenges of the project and the massive quantity of funds that can be at stake.

As we get towards terminal Total Difficulty There is some uncertainty regarding the exact date of The Merge (TTD). As of now it is unclear what the exact date and the exact time that will be the subject of The Merge may be interpreted in various ways. The excitement of The Merge, on the contrary, hasn't been slowed down in any way by this change. It shouldn't come as an unsurprising fact the fact that Ethereum engineers, such as Tim Beiko, were optimistic and thrilled about the results that was this Bellatrix update. With it is clear that the Bellatrix update is over, there is another hurdle to be cleared prior to The Merge can be considered successful. The developers working on the project revealed their enthusiasm for the project last week, however they made sure to that they were focusing on the importance to The Merge because a lot is riding on the results.

The shift between Proof-of-Work (PoW) towards Proof-of-Stake (PoS) that is predicted to reduce the use of energy to 99 percent is the sensible future step in the evolution of Ethereum and makes sense strategic. Since the cost for energy is continuing to increase across the world This change isn't only something that's been long-awaited, but also one which is highly valued. But, the move to Proof-of-Stake is not without controversy. There are legitimate arguments to be presented against Proof-of-Stake and the most popular of which is that it may result in an increase of wealth concentration in the comparatively small number of validators. Additionally, it can make it harder people who are new to the system.

Many people believe that the benefits could surpass the drawbacks, specifically regarding the security of blockchains and the potential for its growth over time. People who favor the change to Proof-of-Stake assert that it will improve

Ethereum's decentralization and distributed nature. According to Buterin the reason for the upgrade for Ethereum is to allow anyone to establish an Ethereum validator without the same amount of technical or financial investment that mining requires. This is extremely important for the foundation of Ethereum and would greatly contribute to making sure the longevity and viability of the blockchain in the event that it was implemented. One expert is of the opinion that if the transfer is successful, it could allow for the development of Ethereum and will increase the number of applications that can be made by it in the future.

What exactly is "The Merge" entail for those who have already registered for Coinbase?

We expect that the trading of the entire range of our centralized trading platforms is not affected at all. Everyone must be vigilant for any frauds in any major event such as The Merge, which is happening right now. In the transition period the assets stored on Coinbase will be safe and secure. There is nothing you must take yourself to upgrade the security of your Ethereum.

I'm trying to prepare in time for The Merge, but I'm not sure of the steps that I should follow.

The simple answer to this question is not. There is nothing for users or holders of ETH to take to safeguard their wallets and money after entering , and throughout the entire duration period of The Merge.

What price will Ethereum change due to The Merge?

As of now, no one is able to predict what's going to happen. The speculation of traders is that prices could go any direction, which implies that certain traders are anticipating an increase in prices and others believe that the reverse could happen. Even when taking into consideration possible deflationary policy and increased activity by developers with blockchains, it's obvious that the outlook for ETH after the Merge isn't exactly crystal-clear. This is true even though it is evident that the outlook for the market isn't particularly clear.

If ETH burning, as well as the network's activity is considered the likelihood is for the probability that ETH issue would decrease by 90%, and it is possible that the total token supply could fall. But, it's not clear how much, or if the reality of this is already being reflected in the cost of Ethereum. Additionally that, if the operating miners take a Proof-of-Work chain from the rest of Ethereum's mainnet This could make things significantly more complicated.

Where can I place the money on my Ethereum or how should I go about it?

Coinbase allows you to take part with Ethereum betting. Our customers have the chance to put your Ethereum (ETH) through Coinbase to earn rewards and there's no minimum amount of ETH you need to put on Coinbase. It is possible to stake your ETH is converted into the cryptocurrency ETH2 after you have staked it. As of now, US users who stake at minimum 100 dollars worth of Ethereum will receive a bonus of 10% (up 30$) in addition to earning 3.25 percent APY for the entire amount of ETH they staked (Terms apply, first stakes only). This bonus is added to the 3.25 percent APY they receive on all their staked Ethereum. It is not applicable for New York or Hawaii).

Continue reading

Is there a possibility of a seamless transition from ETH 2 automatically?

Is there a brand new currency in conjunction to Ethereum 2.0?

After the Merge is complete and the merger is complete, there will be no difference in ETH 1 and ETH 2 which are now called"the execution layer," and "the consensus layer, respectively. ETH 1 will be referred to by the name "ETH."

Are drivers likely to pay less gasoline as a result of The Merge?

Merge Merge Update is focused on altering the process of consensus (the method that Ethereum confirms transactions) instead of growing or expanding the capacity of Ethereum, so it is unlikely that gas prices will fall in the short term in the wake of the Merge upgrade.

Following the conclusion The Merge, what will be the next steps? The Merge, what further kinds of improvements can be in store for us?

Following "The Merge," there will be a variety of upgrades for the game. These upgrades come with names that rhyme with each other, like the rush purge, verge, and the splurge. The surge will comprise the Shanghai upgrade which will also include the ability withdrawal of staked ETH and sharding which is designed to increase the capacity by distributing the load of processing and managing huge amounts of data across the all-encompassing network. Both of these capabilities are part of the increase. A significant step toward greater centralization The Verge will soon allow people to join as network validationators without the need to store large amounts of information. Purge is to eliminate old network history, and the term "splurge" refers to fine-tuning procedures that preceded it.

What is The Merge plan to solve the problem of scaling?

In the wake of the increasing popularity of Layer 2 scaling options, the initial plans to address scaling before The Merge through sharding were revised, and sharding was not scheduled to be launched until 2023, at the very time of the earliest.

What's the reason Ethereum changing its consensus mechanism from Proof-of-Stake to Proof-of-Stake?

Ethereum is moving between Proof-of-Work (PoW) into Proof-of-Stake (PoS) in order to increase the security of the blockchain as well as reduce its energy consumption and lower barriers to entry by reducing the amount of hardware required and also build the infrastructure to improve its capacity.

What is the primary distinction between the Proof of-Work and Proof-of-Stake protocols?

In order to validate transactions Proof-of Work uses the power of computers spread across an uncentralized network. In contrast Proof-of-Stake relies on authenticators who put their money into the token, in this instance, ETH, in exchange in exchange for the possibility of updating Blockchains with latest valid transaction and also earn new created tokens.

Rewards for Staking ETH

3.25%\sAPY*

Ethereum on the social media of today on the day of the closing day and an hour and

On the 14th of September 2022, 2022 The cryptocurrency Ethereum was mentioned in 366,296 of the 1,980,896 social media posts across Twitter in addition to Reddit. Ethereum is currently ranked 2nd in terms of most mentioned and active users according to the number of posts it has collected and has a total of 220,589 users commenting on it.

The Ethereum Merge is in full swing in the present moment.

This week was the week that the upgrade for Ethereum was made available via the upgrade of Bellatrix. Additionally, we'll recap all of the news reports that were reported in the past week, answer the questions you've been asking the most about The Merge Ethereum Merge, as well as provide an article recommendation to ensure that you can be able to learn the more details regarding The Merge.

We examine the execution risks which should be considered in the context of The Merge and investigate the price movements of previous BTC halves to gain insight into what may happen.

Ethereum should prioritise privacy, decentralization, and being truly neutral.

This page gives a brief summary of the latest episode of the Coinbase's Around The Block podcast in which Coinbase CEO Brian Armstrong and Ethereum co-founder Vitalik Buterin discuss decentralization, privacy, and a neutral Ethereum. The show featured host Viktor Bunin.

A look at the Evolution of Ethereum and Investing at a crucial crossroads

Prior to the Merge takes place, we will review both the bear as well as the bull arguments in favor of ETH and debunk some of the misconceptions that have been floating around regarding what the new update might mean for the network and users.

It is believed that the Ethereum Merge is in the works Here's the information you need to know

It is expected that Ethereum will be converted towards Proof-of-Stake (PoS) at or about September 15 2022. The move will render Ethereum safer, require less energy and will better suited to the implementation of new scaling techniques.

There will be a separate discussion of the latest eth2 updates available. We will discuss what merging does and doesn't do as well as the various elements that will happen during the lead-up to the merger and how the rewards are changed following the merger is completed. Concerns over the growing DAG dimension on the internet, as well as details regarding the upgrade post-Merge are just a few of the updates.

Vitalik Buterin, the creator of Ethereum has stated that, following The Merge, Ethereum would undergo several upgrades that have rhyming names like"simple, "surge," "verge," "purge," and "splurge."

The crypto market finally got some respite at the beginning of the week following through a series of weeks of bad news. The recent increase to the cost of Ethereum (ETH) began on Friday when the makers of Ethereum came to a common understanding on a revised timeline for September's much-anticipated "merge" update.

The Merge what does It mean to Ethereum Ethics?

If compared with PoW, Ethereum's switch to PoS will bring about a reduction in energy consumption of approximately 99.95 percent. In a time where the cost of energy is rising across the world and this change will be an improvement which will be highly appreciated.

The scenario can be summarized as follows: It is a difficult year for crypto fans and investors all over the world. Bitcoin is, without doubt, the most valuable cryptocurrency has seen its value drop in the range of 70 percent or more since hitting its record-breaking high nearly one year earlier. Similar can be said about ether, the second largest currency and utilizes the blockchain Ethereum employs.

But, crypto fans have discussed a complicated program update to Ethereum network known by"merge. "merge" for a few months in the past.

According to the non-profit group which is the main force behind the network the merger will simply bring the infrastructure that is the basis of Ethereum in a direction that is more eco-friendly by reducing the carbon footprint of Ethereum by nearly 99%. This is the simple explanation, however the process of actually achieving it took many years of research and testing. It was not clear what the final result was going to be because, like to other developments in the field of cryptography, no similar thing had been attempted before.

It appears that the merger was successful to now without difficulties.

Here's the information you should be aware of:

When you hear people criticizing cryptocurrency asserting that it uses more energy than the entire nation of Argentina such as or comparing bitcoin's footprint to that of all fridge that is in United States combined, they refer to the global computer community that is required to validate transactions using"proof-of-work" protocols "proof-of-work" process.

The Proof-of-Work consensus mechanism is used that is used to verify digital currencies like Bitcoin and Ethereum. It requires powerful computers in order to "mine" fresh coins and validate transactions through a global computer network that is decentralized. Up until recently it was the case that both Ethereum and bitcoin worked by relying on proof-of-work. (I know it could appear to be Science Fiction, yet the more detailed explanation is sure to put anyone to sleep, so I'll keep it as is.) Simply put the proof-of-work algorithm is a disaster for the environment, and extremely negative for the public image of the cryptocurrency industry.

The long-awaited integration has placed Ethereum on the "proof-of-stake" transaction verification method which is energy efficient and requires less electricity.

What do we next?

If the merger is successful without a hitch, then the Ethereum network is the home of all the community members of Non-Future Tokens (NFTs) and NFTs, will continue to function in the same way that it has been in the past, and

153

consume a substantial amount less power and, as per the those who advocate for the technology, increasing the level of security.

Is it possible that Bitcoin ever adopt this approach?

It's not even possible. There are a number of philosophical disagreements in the field of cryptocurrency on the use of the technology behind it.

As per Laura Shin, who is the host for"Unchained," the "Unchained" podcast Ethereum is and Bitcoin have "quite completely distinct culture." "Bitcoiners consider proof-of-work to be more effective in protecting the network" even though it's theoretically possible for Bitcoin to modify its design and structure, as Ethereum has recently demonstrated.

The revolutionary update removes any need to have miners who previously were responsible for driving blockchain technology and promises major enhancements to the environment.

After years of work and delays, the huge upgrade of Ethereum known as the Merge is finally taking place. The Merge has the effect of shifting the electronic machinery at the center of the second-largest cryptocurrency according to market value to one that is much better in energy usage.

Moving from one method of running a blockchain, referred to as proof-of-work, and switching to a different method, known as proof of stake it was not an easy process and required a significant amount of effort. "The metaphor I employ is the idea of removing the engine from a car moving," said Justin Drake an analyst at the not-for-profit Ethereum Foundation who talked to CoinDesk prior to when the Merge was completed. Drake was talking about the process of switching off engines from motor vehicle.

Special Information on CoinDesk Concerning the Ethereum Merge

The potential return is to be massive. It is predicted that Ethereum will require approximately 99.9 per cent less power. According to one estimate it's similar to the case if Finland had abruptly shut down its power system.

The people who created Ethereum think that this new version will ensure that the network is secure and also more scalable. The Ethereum network is already an estimated $60 billion of ecosystem which includes lending companies, non-fungible-token (NFT) markets as well as other applications.

Learn more about what an successful Ethereum upgrade might be like by reading The Merge Monitoring Guide.

More than 41,000 viewers were watching the "Ethereum Mainnet Merge Viewing Party" Livestream on YouTube at the time that the Merge officially started in the 6th minute of 6.43 a.m. UTC. They waited in anticipation for crucial numbers to show up to indicate that Ethereum's core systems had not been compromised.

After what seemed to be an interminable time it was announced that the Merge was officially completed, which meant that it was an accomplishment. In the background of the Merge was not much of a change in cost in Ethereum (ETH) which is worth around $200 billion right now which makes it the second largest cryptocurrency behind Bitcoin (BTC).

The cryptocurrency community's enthusiasts, investors and skeptical people have been keeping an eye to the development due to the impact it could affect the wider blockchain industry. The update will lessen the dependency of the network on the intensive energy use of cryptocurrency mining.

Mark Cuban, investor and billionaire owner of the Dallas Mavericks basketball team, stated to CoinDesk that he'd be "watching the Mergewith interest just as everyone else." He said it could make ETH the network's native currency, deflationary. Cuban has also been the owner of the Dallas Mavericks basketball team.

The notion that Ethereum could one day shift to a consensus model based on proof-of-stake was in place from the very starting of the Ethereum project. However, the transition was a technical challenge and was risky that people were skeptical about whether it would ever happen in any way.

Drake expressed his feelings in a statement, "There's a part of me that isn't aware the fact that this really is happening." "You already know that I've trained myself to expect it to occur in the near future, and I'm sort of denial about it," the person said. "I'm partly denial."

The challenge with the update was complicated by the fact that it was possibly one of the biggest open source software projects that has ever been attempted. This meant it required collaboration among teams of dozens and hundreds of researchers or developers as well as volunteers.

Interview in an interview with CoinDesk, Tim Beiko, an engineer with the Ethereum Foundation who played an integral role in the execution of the update stated, "I think the Merge will really encourage individuals who were interested in Ethereum however, and were skeptical about the environmental impact to try it out and play with the technology."

Chapter 9: Goodbye, Miners

The idea of decentralized ledger was first made available to the world at the time of 2008 through Bitcoin. Decentralized ledgers are an permanent record of transactions that computers from all over the world can be able to read, edit and trust without the need of third parties to serve as intermediaries.

Ethereum released in 2015, enhanced the fundamental concepts of Bitcoin by creating smart contracts. Smart contracts are computer-based programs that work with the blockchain similar to that of a supercomputer to store data on its network. This was the most significant part of the decentralized funding (DeFi) along with Non-fungible Tokens (NFTs) which are the principal drivers of the most recent increase in prices for cryptocurrency.

The Ethereum proof-of-work system has been taken off the market in light the merger known as The Merge. In this system the cryptocurrency miners competed in order to upload transactions to Ethereum's ledger and earn prizes for working through cryptographic challenges.

The majority of the cryptocurrency mining currently occurs in what is commonly described as "farms," but which might be more precisely described as factories. Imagine massive warehouses full of computers that are placed over one another like books on shelves in a library at a university. Every machine is overloaded that it's extremely hot as it struggles to generate bitcoin.

This method, which was developed by Bitcoin and Bitcoin, has made Ethereum to consume such a large amount of energy, and is the reason for promoting the image associated with the industry of blockchain as a environmental threat. Bitcoin was the initial cryptocurrency that adopted this method.

A couple of months ago my daughter and I engaged in a discussion about NFTs, or non-fungible currencies (NFTs)," recounted Ben Edgington, a product manager in ConsenSys which is an Ethereum company that conducts research and develop firm. "While we were having dinner, I accidentally spoke about some NFT initiatives. She was screaming at me: 'How do you cause seas to be destroyed by this kind of nonsense? " It's a horrible scenario. It's difficult for me to comprehend that this is the way you earn your living.'"

Edgington who began his career as a researcher in the area of climate science prior to moving into cryptography, was able to see the place where his daughter came from. He observed, "Whether you think it's correct or not she's been absorbed by a toxic environmental narrative." "I think it's difficult to argue against'stickers for adults' that produce in some estimates an equidistant megaton of carbon dioxide] in a each week." claimed the speaker.

Hello, stakers!

The brand new protocol for Ethereum called proof-of stake removes mining.

Validators are people are those who "stake" at minimum 32 ETH by sending them to an account that is on the Ethereum network, where they can't be bought or traded. Miners are gradually being eliminated to be replaced by validators.

The ETH tokens that have been staked work as lottery tickets: The more ETH an authenticator wagers the higher the chance that some of the tickets drawn will win, thus giving it the right to issue an "block" that contains transactions on the ledger of digital transactions that Ethereum employs.

Beacon Chain was Ethereum's proof-of-stake network, established in the year 2020. Prior to the Merge it served as a place to wait for validators to be ready for the switch. In order for Ethereum to transition from proof-ofwork to proof-of stake The Beacon Chain has to be integrated into the Ethereum main Ethereum network.

According to Beiko that amount of energy transactions that use proof-of-stake employ is "not even an error of rounding" in relation to their impact on the environment.

He explained that using proof of stake was like operating an application an MacBook. "It's the same as operating Slack. It's similar to using Google Chrome or Netflix. In order to function the way it should, your MacBook must be connected into the power supply and run using energy. Yet, no one gives a second thought to the impact the use of Slack impacts the environment, don't you think?

Edgington spoke of the positive impact the Merge remodel has on ecosystem as the benefit that he personally is most excited about. "You know that it gives me an immense amount of joy to think that in the future I'll be able to look back and say I was participant in the project that eliminated 1 megaton worth of carbon out of the atmosphere each week. He added, "That is something that significantly affects not only my family, but also other people."

New motivating factors

It is believed that the Ethereum network is better understood as a nation-state , not an individual piece of open-source software. A nation-state is a kind in living being that is brought together when a set of computers communicates using the same language, and adheres to the same set rules.

This brand new mechanism for Ethereum offers a brand new incentive for people who run these machines to follow the established rules and protects it from unintentional alteration that could take place.

163

"Proof-of-work is a technique that allows you to use physical resources and turn them into security for your network. " Beiko clarified that in order to secure your network requires an increase in the amount of the physical resources. When it comes to proof of stake how we deal with it is to convert the financial resources into security.

While Ethereum was home to thousands of individual miners operating and safeguarding its proof-ofwork network however, only three mining pools accounted for the majority in the hash rates of its network. Hash rates are a measure of the processing capability for all the miners.

It is possible for a tiny group of the largest mining companies on Ethereum to conduct what's called an attack of 51% in the event that they were able to coordinate their efforts to get the majority of the network's hash rates. This could have made it more difficult even for others to make updates to the ledger.

The control over the entire network depends on ETH that is staked not the amount of energy used by every participant in the algorithm of proof-of-stake. People who support the use of proof of stake argue that this makes attacks more costly and ultimately ineffective because attackers are at risk of being able to see the amount they have staked reduced due to their efforts to undermine the network.

The excitement surrounding proof-of-stake crypto isn't shared with everyone. For instance, with Bitcoin there's no evidence that the proof-of-work method is ever going to be abandoned although its advocates insist that it's the best secure and battle-tested method.

Even though control over the Ethereum network will no longer lie within the control of just a few mining syndicates that are publicly listed However, opposition groups continue to argue that the power players currently in place are going to be replaced by new players.

In Ethereum's proof-ofstake chain, Lido, which can be considered an authenticator collective run by the community has greater than 30% stake. A further 30 percent of the stake is shared from Coinbase, Kraken, and Binance that are the three most popular cryptocurrency exchanges.

Chandler Guo, a prominent crypto miner, revealed prior towards the Merge that he will create a fork of Ethereum's original proof-of-work blockchain. It would be a replica of Ethereum's blockchain which functions with the old mining mechanism. The announcement was made in the wake of the doubts expressed by Chandler Guo regarding the proof-of-stake.

Guo's "ETHPOW" project, as well as others similar to it have earned moderate acclaim in certain parts of the cryptocurrency industry however the primary Ethereum developers Ethereum have generally dismissed proof-ofwork forks as a scam and a sideshow.

The Merge is bought and sold

Since mid-July the Merge is a major topic of debate in the market for cryptocurrency. Market participants initially believed that the Merge was an opportunity to trigger an increase in cost of ETH. In the wake of the catastrophe that hit markets for digital assets at the beginning of the year, trading for ETH options started pricing on gains that were made following the Merge that gave investors the much-needed relief.

Traders attempted to secure value from the possible airdrop of the brand newly created "ETHPOW" token, as the possibility of a fork in Ethereum's Ethereum blockchain by angry crypto miners caused a wave of activity. This particular surge was brought on by the possibility of a fork on Ethereum's Ethereum blockchain by crypto miners.

In the majority of cases, it isn't possible to forecast without absolute certainty how markets will react to conclusion of a successful Merge. Because the update is on the list for Ethereum since the platform's creation it is possible for the markets to have priced it in, at a minimum, to a significant degree.

Kevin Zhou, an analyst at Galois Capital, was quoted as saying "I believe that if someone had been asking me about three weeks ago I would have told you that it's not just cost-effective, but also that it's over cost-in." It's now about 70-30 split in the market favor that this is a good occasion for ETH.

Chapter 10: The Ethereum Merge

What exactly is The Merge?

The Merge upgrade marks Ethereum's long-awaited transition from its "Proof-of-Work" consensus system to an "Proof-of-Stake" technology. The Merge is an in-between process that has been dubbed"the Bellatrix & Paris upgrades. The Merge was officially launched by Bellatrix that occurred on the 6th of September 2022 at 11:34.47am UTC. Bellatrix is an enhancement to the network consensus layer. Bellatrix can be followed by Paris which will become an execution layer which moves Ethereum from proof of-work to proof of stake. Paris is triggered by a specific total Difficulty threshold, also known as"the Terminal Total Difficulty (TTD) which determines the day of Paris TBD because it depends heavily on the proof of work hash rate. TTD is the total difficulty threshold required for the block that was mined the most recently in Ethereum. Also, TTD indicates the set of of hashes remaining to be mined before Proof-of-Stake officially takes over.

What happens next after The Merge?

When it is completed, the running Beacon Chain will take over the task of verifying new transactions with the Proof-of-Stake mechanism, while Ethereum's standard Proof-of-Work mechanism will disappear for the rest of time. So far Validators have already placed more than 13 million Ethereum through Beacon Chain. Beacon Chain. The mainnet (the principal system of Ethereum blockchain) is linked to Beacon Chain, the mainnet will be joined with Beacon Chain, the whole transaction history of Ethereum which includes every smart contract, transaction and balance since July 2015 will be joined.

What is the reason The Merge important?

The Merge has been six years in the making and considered by many to be an important moment within the history of cryptocurrency due to the potential consequences both in terms of the material and the philosophical. The milestone may boost market confidence and give investors needed optimism following months of volatility in the market caused by various factors, including inflation and increasing interest rates. According to one reviewer Ethereum's Merge "will prove that a decentralized , permissionless network can be run in a way that is energy efficient." Additionally, a merger such as this is an uncommon event in crypto, and is likely to never happen again.

What will The Merge affect me and my Eth?

The answer isn't very much. The Merge is expected to be be seamless from a user's perspective. Coinbase temporarily suspends certain withdrawals and deposits in order to prevent any risk including the newest Ethereum (ETH) ERC-20 Polygon (MATIC), Polygon (MATIC), Optimism (OP), PWETH, PUSDC and PMATIC tokens. we don't anticipate any other currencies or networks being affected. For those who put their money into ETH the balances won't be released and neither will they be available to sell or transfer as soon as they have completed The Merge. The ETH that was staked is expected to be released and accessible after the Ethereum protocol has completed its enhancements. According to current estimates, the date for upgrade to be completed are in the early 2023 timeframe. Keep an eye on the calendar!

EXPECTED DATE FOR MERGE

The exact Merge date is very unclear as we get towards terminal Tissue Difficulty (TTD) (TTD). As of now, the exact date and timing that The Merge will take place The Merge is anyone's estimate. However, the enthusiasm over The Merge from continuing to make its way into the brewing process. Ethereum engineers such as Tim Beiko, were understandably happy and overjoyed by the results in this Bellatrix update. With the Bellatrix update safely in the rearview mirror, there's one more hurdle to overcome prior to The Merge is complete. The developers working on the project expressed their enthusiasm in the last week, however, they also highlighted how important The Merge is, considering the risks involved.

Ethereum's move to Proof-of-Stake which is predicted to cut down the energy consumption by 90%, makes sense as a move for Ethereum. This change is not just widely anticipated, but is also positive as the price of energy continues to rise all over the world. However, the transition to Proof-of-Stake has its critics. There are strong arguments in support of the concept of Proof-of-Stake. For instance, it is believed that it can lead to a massive accumulation of wealth between a handful of people who are able to verify it and create a barrier to entry impossibly high. Many believe that the benefits eventually surpass the negatives, especially regarding the security of blockchain and the ability to scale it over time. Some advocates claim that this shift to Proof-of-Stake can make Ethereum more robust and less centralized.

According to Buterin Ethereum's latest update, it aims to let anyone set up the Ethereum validator with the same amount of money and technical know-how required to mine. This is extremely important for Ethereum's core layer and would greatly assist in ensuring the long-term viability and the future for the cryptocurrency. If this is successful, it could, at the very least one commentator suggested that it will open the doors to Ethereum and extend its use scenarios for the near future.

What will The Merge affect current Coinbase customers?

We do not anticipate any impact on the trading process across our central trading platforms. If there is a significant event, like The Merge, all users should be alert for any fraud. Your coins are safe with Coinbase are secured and safe throughout the period of transition, and there's nothing needed on your behalf to upgrade your Ethereum.

After the Merge is completed You will find your staked the ETH (ETH2) balance on Your Ethereum (ETH) bank account. For your information that there isn't an token ETH2. The ticker on Coinbase which indicates staked ETH. It will not be used after the Merge. For more information on the implications The Merge will mean to all of our Coinbase customers, read our complete blog post on what to expect in the weeks leading up to and following The Merge.

What do I need to do in preparation to be ready for The Merge?

The answer is simple: No. As holders and users of ETH it is not an requirement to protect your wallet or assets before or through The Merge.

What can The Merge influence Ethereum's price?

It's the only way to know. The market has speculated that prices will be both ways, i.e. certain people are expecting that prices will rise, but others believe that the opposite will take place. It is clear that the forecast for market prices for ETH after Merge isn't very precise considering the possibility of a deflationary policy and increased developer activity in the blockchain. ETH is expected to fall by 90 percent and the overall supply could be reduced in the event that burning and network activity are considered. It's not clear what percentage or if the fact is already integrated into Ethereum. In addition, it could be more complicated if miners currently branch out a Proof-of-Work chain away from Ethereum's mainnet.

Coinbase toInstitutional Research looked into both the bear and bull arguments for Ethereum at this pivotal intersection for blockchain technology.

How can I invest my Ethereum?

You can stake Ethereum through Coinbase. Coinbase allows our customers the opportunity to stake their ETH in exchange for rewards and rewards, and the stake isn't capped at amount of ETH needed to stake Coinbase. If the ETH you have staked it is converted to the ETH2. At present, US users who stake at least $100 of ETH can earn an additional 10% (up 30$) in addition to earning 3.25 percent APY on all staked ETH (Terms apply, first stake only. Not valid in NY and the HI).

The cost of ETH2 is the same as ETH. After the update of Ethereum is complete, Ethereum network is completed the two coins ETH as well as ETH2 will be combined to create one coin. Join an Coinbase account to get up to 3.25 percent APY each time you put your money into Ethereum. Staking options and availability of assets depend on the location.

Does ETH automatically turn into the ETH 2. Is Ethereum 2.0 come out as a brand new cryptocurrency?

When the Merge is completed There will be no difference from ETH 1 and ETH 2 and has been named the execution layer and the consensus layer as well.

There will instead one Ethereum for the foreseeable future. There a misconception to believe that Ethereum 2.0 can be described as a brand new currency or an asset. It's not. your existing ETH will function exactly the same way it did previously. It won't be affected through The Merge.

Does The Merge lower gas fees?

In the short term, The Merge is not anticipated to lower gas costs as the update concerns changes to the mechanism for consensus (the method by which Ethereum confirms transactions) instead of the expansion or increase of its capacity.

Also, the Merge will end PoW and move from the Blockchain to PoS however, this move will not be a problem for the capacity of the network. Because gas prices result of demand in comparison to. capacity of the network, rather than the way that blockchains verify transactions This means that The Merge doesn't directly impact the cost of gas.

What additional enhancements are planned after The Merge is complete?

There are many more updates following The Merge. The rhyming upgrades are called The Verge and Rush purge and splurge. The surge will come with the Shanghai upgrade which will provide the capability for withdrawals of staked ETH as well as sharding, that aims to improve capacity by spreading the burden of processing and managing huge amounts of data over a entire network. The verge will enable consumers to become network validaters without needing to maintain large amounts of data which is an important move towards a greater decentralization. The purge will aim to erase any history from the network while the splurge is referring to the refinement of its previous methods.

Although The Merge marks a major achievement in the Ethereum blockchain but it's certainly not the only major occasion. There are likely to be many more developments to come and plenty to be excited about post-ETH Merge.

How will scaleability be managed how will scalability be handled The Merge?

The initial goals of tackling scaling, via sharding prior to The Merge were changed owing to the emergence in layer two scaling solutions with sharding set to be launched in 2023.

Sharding lets a database split horizontally in order to spread the burden. This means that it helps in spreading the enormous burden of data processing over the entire network. This should help reduce the amount of traffic and increase the processing time.

What is the reason for Ethereum changing to Proof-ofStake?

Ethereum is moving from Proof-of-Work (PoW) in favor of Proof-ofStake (PoS) as a way to increase the security of the blockchain and making it less energy-intensive and lower the barriers to entry by reducing the requirements for hardware, and create the infrastructure to increase scaling.

How can you tell the differences between Proof of Work and Proof-of-Stake?

Proof-of-Work makes use of computers on an uncentralized network to verify transactions, whereas Proof of-Stake relies on validaters who contribute their tokens for this particular instance, ETH, as collateral in exchange for the chance to upgrade Blockchains with most recent verified transaction and to earn newly created tokens.

How can the Merge assist you?

The Merge will bring about Ethereum completely rethink its proof of work its energy-intensive system that it currently employs instead of evidence of stake.

In the world of crypto, "staking" refers to the deposit of cryptocurrency into a protocol. Sometimes, it is to earn interest. For example, the creators of terraUSD stablecoin stated to customers 19.5 percent linterest in the staked TerraUSD The possibility was to put $10,000 in and cash out $11,900 over the course of one year (until it went down) (until it exploded).

In other cases, for instance, for instance, a proof-of-stake-based blockchain that is secured by staked bitcoin, it helps to safeguard an entire system. We'll be seeing this in the near future. the more ether that is staked and the more secure the blockchain will be following the Merge.

With the proof of stake is accepted, miners won't longer be required to work on complex cryptographic issues that consume a lot of energy to verify new blocks. Instead, they'll put

Ethereum tokens into an account. Imagine that each token are a lottery ticket. If your token's number is drawn, you have the chance to validate the next block, and earn the rewards that go along with the process.

It's a cost-intensive business. Potential block verifiers -- - who will be known by the name of "validators" as opposed to miners- have to put up a minimum amount that is 32 ether ($52,000) for being legally eligible. The method involves punters putting in cash rather than power to confirm blocks. While a criminal needs 51% of the network's power to take over the proof-of-work system, they'd require 51% of total staked ether to overwhelm the proof-of-stake mechanism. The more ether that is staked, the more secure the network will become when the cost of achieving 51 percent of its capital rises.

Since cryptographic issues won't longer be a part in the Ethereum system the energy consumption will decrease by approximately 99.65 percent, as per the Ethereum Foundation.

What are the risk factors?

The Ethereum Foundation said that PoS can reduce energy use by as much as 99.95 percent

The shift from the Ethereum network to a proof-of-work (PoW) is likely to result in an increase in the number from idle Ether (ETH) mining companies into the market for cryptocurrency. Like Bitcoin, Ethereum has been created on PoW blockchain since the year 2015. The network will be converted into Proof-of-Stake (PoS) once the update is completed that will fundamentally alter the method by which transactions are validated in addition to being added to the Blockchain.

"The proposed consolidation will have a number of possibilities of benefits. If it is implemented correctly the new system will use 99.5 percent less energy as than the current method. Furthermore, once the PoW network ceases to function, Ethereum issuance will fall dramatically, increasing Ethereum's value. Ethereum as a result. Overall, the changes in protocol will increase the value of cryptocurrencies and draw an influx of new individuals and institutional investors." Swarup Gupta, financial services head and lead, ESG, Economist Intelligence (The EIU) said to FE Blockchain.

Additionally, there seems to be an increase in the fear that could result in negative financing. "Traders are attempting to either buy ETH on spot markets or trade or hedge stock in the near future. In this day, customers will be able to acquire a similar amount of additional Ethereum PoW (ETHPOW) tokens that will be unlocked and could be traded on the open market. This excess short-holdings in futures and perpetual contracts could cause the negative impact of funds." Sumit Ghosh, co-founder and CEO of Chingari said.

Additionally, as the popularity of merging, Google has been beginning to show the remaining time to merge based on the present difficulty, hash rate and the difficulty. According to experts in the industry, "There are some little problems that could arise that could be fixed down the way. Validators of proof-of-stake will take on the responsibility to verify the authenticity of all transactions , and also recommending Blocks," Rahul Kapoor, co-founder of CryptoRunners said.

Conclusion

Thank you for taking the time to read to the end of this book. Hopefully it has been informative and will be able to provide you with all the tools you need to reach your goals, no matter what they might be.

The next stage is to provide an announcement about the way you plan to invest your money into Ethereum as well as different digital currency. Although you may think that we're at an end point that isn't the case. The value of the largest digital currencies is expected to increase in the coming year and even years. In the final quarter of 2017 the price could rise by over 10 percent. In a short time the cost could double and then increase. With all the legislatures on the globe approving rules and regulations to impose electronic forms of currency and a corresponding expectation that these financial standards will soon be utilized to pay for labour and goods across the world.

Experts in the area of innovation have stated that the blockchain, the technology that is decentralized and that is used by a lot of electronic currencies is among the most effective advancement since the Internet. The experts know exactly what they're discussing. Anyone with a keen mind would like to invest their money in this kind of venture.

But, even if you are planning to put money into Ethereum it is important to stay clear of risk and invest with care. The slow, but specific financial backing company dominates the market. You must move slowly, buy wisely and keep building your part of Ethereum or any other cryptographic currency regularly. Make sure all your cryptographic currency as safe and secure as you can reasonably expect. If you've invested your money into these currencies, you need to ensure that they are safe.

If you adhere to the guidelines that is provided and invest with care and wisely, there's no doubt that in a few months or even years from now you'll be happy you made the investment in Ethereum which will earn the money.

www.ingramcontent.com/pod-product-compliance
Lightning Source LLC
Chambersburg PA
CBHW071219210326
41597CB00016B/1872